ECO-RENOVATION

ECO-RENOVATION
The Ecological Home Improvement Guide

Edward Harland

with illustrations by Duncan Roberts

CHELSEA GREEN PUBLISHING COMPANY
WHITE RIVER JUNCTION, VERMONT
TOTNES, ENGLAND

This revised edition published in 1999 by
Chelsea Green Publishing Company
P.O. Box 428, 205 Gates-Briggs Building
White River Junction
Vermont 05001
www.chelseagreen.org

U.S. Library of Congress Cataloging-in-Publication data available on request.

First U.S. edition published in 1994 by Chelsea Green. Original edition published in 1993 in the U.K. by Green Books Ltd., Foxhole, Dartington, Totnes, Devon TQ9 6EB, in association with The Ecology Building Society, 18 Station Road, Cross Hills, near Keighley, West Yorkshire BD20 7EH.

Printed by Biddles Ltd
Guildford, Surrey, U.K.

ISBN 1 890132 38 1

CONTENTS

Acknowledgements

Firstly I would like to thank Tony Weekes for his help and encouragement in initiating the project in the first place. I am also grateful to Luise my wife who helped in checking some of the early material; and my daughters Effie and Vivien who put up with reduced family life during the main writing period. Gus Smith also provided some useful comments on the first draft, as did Peter Inch on the Space chapters, and Matthew Hill suggested technical improvements to the Energy chapters at various stages of the writing. Patricia Spallone also responded at short notice to give detailed and useful comments on the Health chapters. Duncan Roberts not only produced some inspired drawings but also developed many of the ideas for the illustrations; and finally I would like to thank John Elford of Green Books for his encouragement, patience and many helpful suggestions at all stages of its writing.

To Ole P. Larsen

INTRODUCTION

THE AIM OF THIS BOOK is to help you to put into practice ideas and approaches to renovating your house in an ecological way. Whether or not you are a committed environmentalist, by following the principles laid out in the following chapters you will be able to:
- Organise the space in your home to better effect
- Save energy (and money)
- Live in a healthier environment
- Choose environmentally friendly materials
- And make a small—but significant—contribution to the well-being of the global environment

As this book will show you, there is no doubt that the nitty-gritty details of our domestic arrangements affect the global ecosystem in quite profound ways. Our existing houses are 'given': we cannot undo any damage to the environment that has already occurred through their construction, improvements or day-to-day use. However, we can begin to make the changes that will lead them from having a basically damaging effect on the environment, to having at worst a neutral and at best a healing effect.

Before we get on to practicalities, we need to take a look at the broader picture. What are the most critical ecological problems at present? The perceived importance of particular problems changes with new scientific information, fashion, political expediency and our own inability to cope with the enormity of the unfolding ecological crisis. Some problems stand out from the rest. Perhaps the population explosion is the parent of them all. Climate change resulting from global warming can be seen largely as a result of consumption patterns in the West; we cannot foresee the final result of these climatic mutations. Our unwillingness to think of the generations to come as we burn up fossil fuels almost as quickly as we can extract them from the ground is another example. We hope that science and technology will come to the rescue. However, as we are finding out, these in themselves bring us knowledge and power, but not wisdom. It is we humans who have to learn to choose which directions in science and which applications of technology will help us out of our present ecological problems. Science and technology have brought us vast numbers of new synthetic chemicals, some in unimaginable quantities. Our careless use of these has led to ozone depletion, air and water pollution and the degradation of soil, on which so much of life depends.

What are the guiding disciplines that should be leading us to solve these problems? At present, outdated economic thinking is still being used as the final arbiter of most policy and action. Our present economic system relies heavily on unsustainable growth, centralised industrial power and opportunism. In the long term, the medium term and increasingly the short term, unless we run our societies on ecological lines nature will impose its own unilateral solutions. The sooner our economic system, which dominates the whole ethos of our society, follows this ecological imperative the sooner our future on this planet will be placed on a firmer footing. Our housing practices are no exception.

Because of the questionable basis of existing economic logic, as outlined above, I have avoided using cost calculations in the book as a general means of deciding on a particular course of action. You, the reader, must make your own judgements in weighing any increased costs against ecological benefit.

Our present problems largely stem from our having become alienated from the natural world on which we depend. Urbanisation and the comforts of modern life cocoon us into a state of false security. Our houses provide us with most of our basic needs for shelter, warmth and protection. Food and water are transported in to complete our basic requirements. If our work is also in our home, we can even survive quite well without going out at all. We can live in a world of TV advertising, computer communication, constant temperatures, instant food and drink, wall-to-wall carpeting, and so on. Yet we are born a part of nature—and we now have the knowledge to re-integrate ourselves into the fabric of the ecosystem that gave us life in the first place. We now need to adapt our homes by applying this knowledge. By learning to relate more directly the ecological events in the world to ourselves and our homes this process can be accelerated. It is clear, in a given situation, that many different ecological solutions are possible. To know that this is true, we need only look at the enormous diversity of nature to see that there are many ways of achieving ecological sustainability.

I have organised the many solutions into four areas: Space, Energy, Health and Materials, which form the main chapters of the book. Within each of these subjects or chapters fall the important principles of ecology: recycling, self-sufficiency, renewability, conservation and efficiency. The essence of ecological thinking is to learn as much as possible from nature and if possible to participate in these natural processes and integrated systems, albeit in a more technical way.

Perhaps the most important principles of all relate to our adaptation to our natural surroundings and the recycling of resources. These principles can be applied directly to our housing and the question of whether to rebuild or renovate. The answer is clearly that renovation is the primary task. New houses only represent less than 1% of the housing stock annually. We simply do not have the resources to replace all the existing houses, even if we wanted to. Although there are some wonderful examples of sophisticated ecological design, most designs for new housing fall far short of this, wasting even more valuable resources in the meantime. Not all houses should be renovated, but in most cases it is clearly better to re-use, adapt or extend an existing structure rather than demolish and re-build.

Now to the practicalities: our very own homes take centre stage. If you find the analysis of environmental problems somewhat daunting, you may like to move straight on to the 'action' parts in each chapter. In a similar vein, a small project successfully completed is infinitely more satisfying—and effective—than a large one abandoned halfway through.

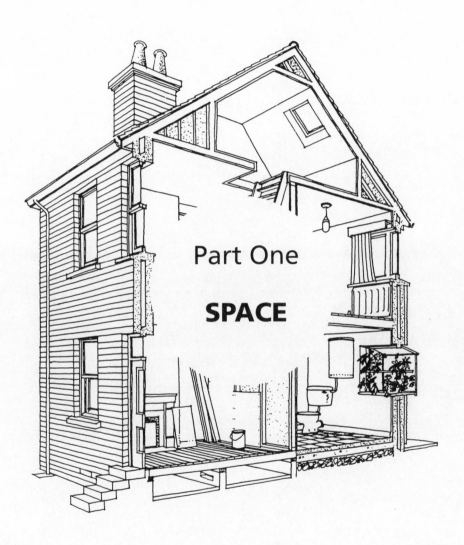

Part One

SPACE

Part one

SPACE

THERE IS MUCH MORE SCOPE than many of us realise to make better ecological use of the space in our homes and gardens. Why is space an ecological issue? It is not usually seen as such. If we think about it for a moment, every bit of space that we have in our homes requires maintenance, heating, lighting, ventilating, painting, and furnishing. Thus in general, given similar construction, a larger space or a larger home will use more resources than a smaller one.

In a wider context, we human beings have appropriated more than our share of space on the planet. We have an ecological responsibility to manage our settlements and housing in such a way as to confine our use of space to reasonable limits. What may seem reasonable to one person, of course, will seems totally unreasonable to another from a different culture. However, without extending the scope of the book to include complex social issues, what I am suggesting here is a heightened awareness of the efficiency with which we use the space available to us. Energy efficiency is a term we are becoming increasingly familiar with as being synonymous with good ecological practice. Space efficiency should be understood similarly: in many ways it is even more basic than the efficient use of energy.

This first part of the book looks at some of the different ways we can improve on this particular type of efficiency, both inside our homes (**INTERNAL SPACE** and **CONVERSIONS**) and outside in our gardens or yards (**EXTERNAL SPACE**). Besides the efficient use of space, some ways of using it more ecologically such as self-sufficiency activities, are also analysed. The final chapter looks at **EXTENSIONS**—ways of extending your home ecologically, should this be necessary.

Home means many different things to different people. It can mean a large house, a semi, a small two-up-two-down, or a high-rise flat—and any of these can be either owned or rented by the occupants. External space can vary from the non-existent through a balcony or concreted yard to a large garden or even a farm. There will be some sections or subsections that you may not find relevant to your particular situation. However the general principles can be modified and applied in all situations.

Internal space

Most of us fill our homes with clutter, and yet we are often aware of the advantages that a less cluttered house would bring, not least that it would look more attractive. What are some of the ways in which we can make the best use of the space in our homes? We can for instance make the most efficient use of circulation space—passageways, landings and hallways—by giving this space a second use or merging it with another area. We can also ensure that every room we have is being used to the maximum extent compatible with the functions or activities of that room. Often, using the same space, we can double up activities which complement each other such as a study-bedroom or a kitchen–dining room. We can also look at the internal arrangements within each room. In some cases there may only be one way of arranging the furniture, but in most there is the possibility of improving the layout to enable better use to be made of the space within. Our living spaces can become more adaptable, as they used to be in the past.

From an ecological point of view, besides looking at the general efficient use of space, which may require us to use smaller areas for some activities, we need to think how we can provide space for self-sufficiency within our homes. All this requires allocating time for careful planning which is essential if we are to optimise our use of space. This section looks at these variables we have to work with and how we can make the most of the technology we have at our disposal. The first issue, occupancy, is not only of ecological relevance but has a wider social significance.

Occupancy

We cannot go very far in a discussion on internal space without looking at the number of people who may be living in a given household. Many houses in Britain are occupied by only one person who has more space than they either want or need. However, there are many with the opposite problem also: of a home that is too small or overcrowded. Ecologically we should be making the most efficient use of housing, and we are far from achieving this.

Is there an optimum number of persons that can be accommodated in any particular house? Perhaps the answer is that most people are aware themselves whether they are cramped for space or have additional capacity. There are a number of ways to alleviate these imbalances, for those that have spare capacity in their homes:
- Sharing the house with someone else
- Offering a spare room to a lodger(s)
- Offering the spare space to an elderly parent or relative, either in the

form of a granny flat or spare room
- Sharing the house with another family, to help share child care and provide playmates or surrogate brothers and sisters, in an age when small families are becoming the norm
- Offering bed and breakfast

For those that cannot afford their own house, sharing the purchase of a property with others in the same situation provides a way forward. In general, cooperation and the sharing of resources is always ecological as well as economical. Learning the discipline and social skills necessary for this is of ecological as well as social importance.

Self-sufficiency

Although the growth in GNP (Gross National Product) desired by governments is based on increasing inputs and outputs, there are many of these that we need to reduce if we are to live more ecologically. Some of these we have traditionally made space for in our homes, such as food preparation and storage; others, such as allocating space for collecting materials and products for recycling, we have not. We should first become conscious of the large number of ways that we can be self-sufficient. The more we can grow, store and prepare our own food the better. Also making, repairing and recycling articles for our needs all helps to reduce unnecessary consumption.

Growing our own food

It may seem difficult to achieve much indoors, but there are some opportunities which are more appropriate inside because of the difference in temperature, especially in winter. Sprouts grown from seeds and pulses, for instance, can be grown almost anywhere that is warm, and these do not even require light. We can also grow herbs on window sills. Mushrooms can be grown in complete darkness in a cellar or outhouse, and yoghurt and similar cultures can be seen as a way of growing fresh food. If you have a conservatory, you may choose to grow plants and vegetables that cannot be grown outside.

Storing food

Storing food to provide for the winter is not as necessary as it was in the past. However we should ensure that any produce we do grow is kept in the best condition possible. Apples from an apple tree or potatoes from an allotment require space allocated in the coolest location possible. Systems for ensuring that no food is kept longer than its recommended eat-by date are important.

Processing our own food

We are buying more processed food than ever at present. Often this means the centralisation of production with all the inefficiencies of packaging and transport that result. Preparing food ourselves, shortly before eating it, is also healthier. A well-designed food preparation worktop is the heart of a good kitchen.

Making, repairing and maintaining

Looking after our possessions prolongs their life and reduces the need for replacements. Suitable workspaces and kits for the repair of clothes, furniture, toys and appliances etc. greatly facilitates these activities. Good maintenance habits are all ecological and should be given the necessary priority.

Recycling

There is almost nothing that cannot be recycled. However it is an important principle that these materials need to be separately categorised. This requires space for a storage system of drawers, cupboards or containers. An ecological household will put as much energy into the proper disposal of goods and materials as it does into procuring them in the first place. Even the recycling of water, heat and air is possible to some degree and requires the allocation of space for the necessary equipment.

Allocation of space for recycling.

There are many more ways in which we can be self-sufficient: in education, entertainment and health, for instance. The important point is to give priority to the activities which reduce our unecological levels of consumption.

Small spaces, small homes

Smallness in the past has often been equated with poverty. However people with plenty of money spend their holidays in very cramped conditions on yachts, in caravans or in tents and can enjoy themselves. In many cases it seems cosier. What is noticeable now, especially in large cities like London and New York, is the partly economic and partly fashion-driven trend to make the very most of a small room or flat. The one-room flat becomes a totally designed space with all the necessary functions slotted in by using imaginatively designed fitted furniture and fold-away systems. Inspiration for these designs has often been gained from the interiors of train sleepers, boats or even the space shuttle.

Why do some of us think we need such large houses? It can be a desire for status or somewhere to set off or store an excess of possessions. Alternatively, a large house can be used as a way of remaining aloof or private from others. Whatever the reasoning, we should remember when we are thinking of moving house that having a small, appropriately-sized home has considerable ecological benefits.

Multi-purpose rooms

Small space living necessitates using the same space for different activities—one of the most obvious ways of making the best use of space within your home. Of course most people do this anyway with certain rooms, but it was much more common in the past.

All housing started as a one-room dwelling space and this can still be seen in many parts of the world. However our rooms have become gradually more specialised along with the increased specialisation of our lives.

The following are the commonest activities that we perform in our homes:

Multi-purpose use of space.

- Sleeping
- Food storing and preparing
- Eating
- Washing our bodies
- Excreting
- Washing clothes
- Relaxing—reading, sitting
- Playing games
- Craft working or DIY
- Cleaning
- Meditating
- Growing plants
- Entertaining friends
- Office working
- Watching TV
- Playing music
- Exercising
- Writing & homework

Some opulent houses have separate rooms devoted to each of these activities, often more for show than for convenience. Each activity can even be sub-divided to spawn dressing rooms and breakfast rooms etc., but most of us combine some of the activities and fit them to the rooms we have available. However, we still often feel that to have more rooms would be an advantage. We need to balance up the seeming convenience of a separate room for each activity with the wastefulness of having expensively equipped accommodation idle for much of the time.

With all the ideas and technology available today, it is possible to achieve ingenious arrangements with fold-away beds, revolving furniture, the use of blinds, curtains, and special lighting, etc. This can be an exciting process to embark on—if you need inspiration, some magazines have sections specialising in rooms where spaces have been used to maximum effect. The treatment can seem expensive in time and money but is usually considerably less costly than a conversion or extension.

You can think which activities already occupy the same space in your home and then, by looking at the list above, which further combinations you might be able to employ in order to make space for self-sufficiency or some of your more pressing requirements. It is in the detailed design of these rooms that success depends; if you feel unsure in this area, call in a designer who can help you work out an attractive arrangement.

Space for a retreat or meditation room

One example of combining activities which does not involve extra furniture is the idea of combining a room as a meditation room or retreat. Here the

Bathroom retreat.

idea is to choose a room in the house which you can strip of all clutter and have only things in the room which you find harmonious—simplicity is the key. I am choosing to emphasise this as it brings together the ecology of space within our homes, our physical health and psychological well-being. Some people choose the bedroom, others the bathroom or the attic. This ensures that there is at least one room in the house free from clutter which allows us to clear our mind and focus on whatever we choose, be it meditation, positive visualisations, empowering ourselves, exercise or simple relaxation.

Storage

Most of our houses become filled with too many possessions as we steadily acquire more gadgets, memorabilia, books and magazines, furniture, different kinds of tools, clothes etc. Before we start deciding what storage we require we should decide what are the categories that are the most essential to us. It makes sense to organise the possessions we do decide to keep in appropriate storage facilities, in such a way that they can be easily found when required. Good storage facilities can also give a house a feeling of simplicity. There are many other advantages: it keeps things in order, accessible and in good condition so that they last longer. It can also create more space where we want it. When relieved of clutter we can enjoy the true spatial nature of a room.

Types of storage

What are some of the different approaches to storage we can adopt? These can be grouped into the following categories:

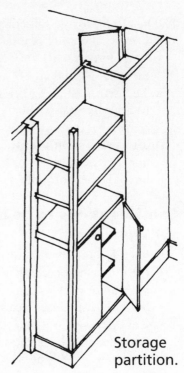

- Free-standing furniture
- Built-in or fitted cupboards
- Modular systems—fitted or free standing
- Rooms given over entirely to storage
- Storage partitions

Within these storage systems themselves, there are many ways of arranging the support and methods of containment:

- Shelving—adjustable or fixed
- Hanging from—hooks, runners and rails
- Containing in—drawers, boxes, lockers, tubes, jars, baskets, trays, box files, tins, and industrial containers.

Storage partition.

It is free-standing furniture and modular units that make the most impact, in our rooms, whereas built-in units have the effect of streamlining the space in question. Our choice of the system for storage thus has an important effect on the way we can utilise the floor area and the impression that is given by any particular room.

Where to locate storage

Where to locate our storage systems or units is perhaps the most important decision of all. The following are the main considerations:

- Look at the entire house, decide which are the least usable spaces, and use these first for storage.
- Consider whether there are ways of using the circulation spaces: the corridors, hallways, landings and stair wells.
- See whether there is spare capacity in the roof space.
- Decide which things you want to have available near at hand (the items you need but never find).
- Also think of having storage associated directly with activities.
- Consider using storage on a seasonal basis (i.e. if there are some things that you would like to have available in winter but not in summer and vice versa).
- Or perhaps one smaller room could be given over entirely to storage.

The height of storage (whether we store things on the floor, on a wall or at high level) is also a relevant factor:

FLOOR LEVEL STORAGE: beneath beds, in the eaves of the roof and under the stairs—eased access with wheels, runners, rails and drawers.

WALL STORAGE: in partitions accessible from both sides, in alcoves and hanging systems of shelving or storage units—hidden from view by curtains or blinds.

HIGH LEVEL STORAGE: the space at the top of the stairwell, round the upper circumference of high rooms, and systems hung from ceilings— use of ladders or pulleys can ease access.

Considerations to be born in mind when incorporating storage

- High level storage requires safe access.
- Light in a cupboard or walk-in store is important.
- An attic or roof space is of limited use unless converted.
- Marking containers clearly helps accessibility.
- Consider weight when you are fixing storage that requires taking extra heavy loads. Check the strength of the wall or floor.
- Store heavy items at waist height if possible to ease lifting.

- Light and bulky items can be stored at higher levels.
- If higher shelves are to be accessible from standing position, they need to be shallower than if designed to be accessible with a stool or ladder.

Furnishing

When we are concerned with furnishing our home ecologically, what are the most important considerations? Here we are concerned with furnishing in such a way as to make the very most of the space, with the most efficient use of materials, and at the same time create an attractive atmosphere. These are some of the types of fittings or furnishing that may help:

- Fold-away furniture such as hinged chairs and tables
- Multi-functional furniture such as sofa-beds
- Small furniture such as kneeling stools (T-shaped wooden support)
- Light furniture such as picnic chairs and tables
- Inflatable furniture such as beds and chairs
- Soft furniture such as cushions or beanbags
- Hanging furniture such as hammocks
- Demountable or knockdown items, such as trestles and flatpacks
- Wall-hung furniture such as seating and tables

We can achieve many different effects with these furnishings. Add to these the creative use of curtains, blinds and lightweight partitions and you have all that you need for living a more flexible way of life and indeed easing the often traumatic experience of moving house.

Priorities for action

1. If you have spare capacity in your house, consider ways of accommodating other people in order to make the most of the space.
2. List all the ways that you would like to be more self-sufficient through activities such as growing or recycling, and give spatial priority to these activities.
3. Think which activities in your home might coexist, or even complement each other, so that you can consider designing multi-purpose rooms.
4. Go through your possessions and recycle any items you don't need by selling them, giving them away or taking them to a charity.
5. Think through your storage needs carefully and look at your existing arrangements. See if there are any new ideas you can apply from the storage section.
6. Does your furniture require a rethink? Are there pieces that keep getting in the way? Can these be recycled or replaced with more flexible items?

Conversions

What can you do if you have analysed carefully your use of space and decided that you need more accommodation? You might want more space for a variety of reasons, such as a growing family, the need to work from home, or an elderly relative to be looked after. Assuming that you have taken into account the measures outlined in the previous section to make the very most of your existing arrangement of space, and that moving house is excluded, you next need to think of ways you can rearrange or develop the space contained within your existing external walls and roof. You have three possible options:

- Restructure the existing rooms and circulation space to convert them to a more useful and effective arrangement
- Convert existing uninhabitable space such as an unused roof space or basement
- Extend beyond the limitations of your existing walls and roof

From an ecological perspective you should choose the solution that uses least resources; the options are thus arranged in order of ecological soundness. This section looks at the first two of these options, the third option being dealt with in the EXTENSIONS chapter on page 28.

Restructuring of existing rooms and circulation space

At first sight it may seem that there is no unused or inefficiently employed space in your home. However, careful analysis often shows that you can create a more effective arrangement of the existing space. The following are the main choices available:

Opening up options

- Create one room out of two, if a more open plan arrangement works better.
- Open up a room with adjacent circulation space such as a hallway or corridor to create a greater feeling of space. There may be particular scope around stairs or poorly designed built-in storage cupboards.
- Change the position of a door to make better use of space within a room.

(If you are removing a wall, care is necessary to ensure that any structural function the wall performs is adequately substituted. Obtain professional advice.)

Subdividing options

- Add a platform to a room with a high ceiling, creating additional intermediate floor space for sleeping or working.
- Subdivide a room into two smaller units.
- If you have rooms with high ceilings, it is sometimes possible to obtain three rooms out of two vertically, by gutting the intermediate floor and replacing it with two. (This option will almost certainly require professional assistance.)

Changing use options

- Sometimes it makes sense to change the location of a kitchen, bathroom or toilet. Because of the services involved, this is usually more of an upheaval than simply changing a living room into a bedroom or vice versa. However, you may find you can create more useful space overall by making such changes.

Conversion of existing uninhabitable space

Once you have exhausted the possibilities of restructuring already habitable space, the next option is to convert any uninhabitable space lying within the confines of the existing walls and roof. The following is a list of the main conversions that are commonly carried out:

- Roof space conversion
- Basement conversion
- Integral garage conversion
- Out house conversion
- Top of stairwell conversion

These vary enormously in their complexity and will depend very much on the particular building construction and spatial layout in your home. The main considerations related to each type of conversion are outlined below.

Roof space conversions

A roof space conversion is one where a previously unused roof space is converted to a fully insulated and lined room or rooms with added dormer windows or rooflights to provide the necessary daylight. In most older houses with open roofspaces, this is usually the most productive of all conversions. However there are many modern houses where it is difficult to achieve this, since the roof space is a dense network of trusses. If this is the case, it will be necessary to replace these trusses for full use to be made of the space, and the help of an engineer will almost certainly be needed.

The second problem which frequently occurs with roof space conversions is finding the best position for the stairs. This can be difficult to achieve with certain configurations where headroom in the attic is limited. There will also be a loss of space in the floor immediately below due to the new stairway. If no immediate solution is apparent, try to see what has been achieved in houses of similar design. Drawings showing vertical cross-sections are essential to check the feasibility of any preferred arrangement. Related to this are fire precautions and the means of escape if the attic is to be used as a bedroom. This aspect requires careful thought. Building regulation approval will in any case be required for such a conversion.

In terms of the application of energy conservation, good insulation is the most important. The different ways to insulate a roof space are outlined in the INSULATING chapter of Part Two. It is likely that the ceiling joists will need strengthening to take the added floor weight. Care is necessary in the provision of adequate cross ventilation for warm weather and for airing. The placing of rooflights with ventilating slots on both sides of the roof will resolve this.

Basement or cellar conversion

The next most common conversion is that of converting a basement or cellar. Here the main problem lies with damp. With this type of conversion it is necessary either to hire an architect or builder who has been successful with damp-proofing similar basement accommodation, or to become highly knowledgeable on the subject yourself. There are many new techniques for achieving this besides the well-known ones of tanking or impregnating the walls with damp-proofing compound. Much will depend on the existing foundations, the water table and any form of previous damp-proofing.

It should be remembered that cellars initially were built so that damp could evaporate there, in order that timber constructions at ground level could remain relatively dry and free from the risk of rot. One of the most important parts of the design should be the detailing of any ventilating cavities to protect such timber members that might be vulnerable.

The adequate provision of insulation, ventilation and daylight where possible are important considerations. However, it is useful to know that dry earth is a good insulator, and that insulation is mainly important at ground level and for a metre or so below (see INSULATING chapter). Ventilation can always be provided using cross ventilation, even if to do so requires the introduction of a duct such as a plastic pipe through any obstructing wall or bank of earth. Daylight can be achieved by installing additional windows and possibly by reducing the level of the ground externally.

Integral garage conversions

Not many houses will have this potential, which occurs only in more recently designed housing where an integral garage is included in the ground floor. These conversions do not usually present any particular difficulties, except for the increase in the insulation that is usually required on the external walls and roof.

Outhouse conversions

Outhouses may consist, for example, of outside toilets, old coal sheds, old wash houses or even old stables. In some cases the most sensible thing to do is to rebuild completely, or partly rebuild using as many of the existing materials as possible. If this is the case the project will have more in common with an extension. Attention will have to be given to planning approval, insulation, weatherproofing, services and drainage.

Top of stairwell conversion

Many houses have unused space at the top of the stairs which, if large enough

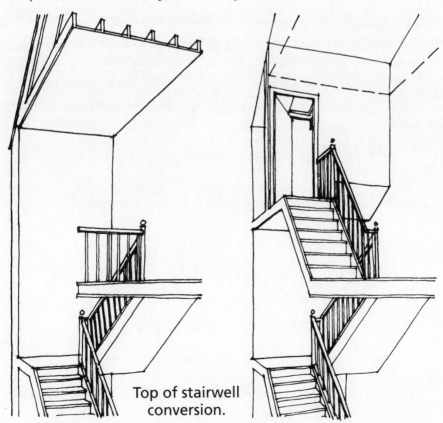

Top of stairwell conversion.

and with high enough ceilings, can be used to create a bathroom, small bedroom or at least additional storage space. To achieve this it is often necessary to break through the ceiling and use the eaves of the roof space directly above. The illustration on p.15 shows one example of such a conversion. A rooflight or dormer window can be incorporated into the roof above. The main problem with this type of conversion is that of incorporating the stairs required within the limited space available. The geometry of the space is all important. As with roof space conversions, cross-sectional drawings help to analyse what is possible.

Priorities for action

1. First, look at the measures outlined in the first chapter INTERNAL SPACE to see if you can make better use of your existing rooms and circulation space.
2. Secondly, look at any rearrangements of space that can be achieved by opening up or subdividing, as described on pages 12-13.
3. Thirdly, look at whether you have any uninhabitable space that could be converted, such as attic or basement space, outhouse or garage, or even at the top of the stairwell.
4. Seek professional advice where it has been recommended, unless you have the expertise yourself. Also refer to the other chapters in the book for advice on energy, health and choice of materials.

External space

Our external space may be a garden, a paved back yard or odd pieces of ground left over between buildings. We think of it in many different ways: as a place to run the dog, an external room to picnic in warm weather, somewhere to sunbathe, a place for the car or to hang out the washing. If we have the space, how many of us see it as an opportunity for regeneration—for recycling our organic matter, providing habitats for birds and animals, storing water and growing food for ourselves?

Our outside area is where we can have fun practising our understanding of ecology. We can, if we feel inclined, experiment with a myriad of plant species, with filtering 'grey' water in reedbeds, with making compost and improving soil fertility, caring for endangered species of plant, animal or insect, or even experimenting with earth walls. The advantage of this outdoor area is that if we treat it with care it provides us with some wonderful opportunities to work with nature. Looking after the garden is usually seen as a leisure activity and it is good to keep it this way: as a means of renewing ourselves.

This section is for those with gardens. If you are without a garden, you may have still have a yard or a balcony. Alternatively, you may have access to an allotment, a friend's or relative's garden, or even to a piece of public land that needs looking after. If you are interested, whatever your circumstances, it should be possible to have access to gardening at some level, even if this means making the very most of your window boxes or tubs of flowers and tomatoes.

Structuring your garden

In all gardens we have to make basic decisions as to how the space in the garden should be allocated. In an ecological garden we can categorise our use of space under three headings, which are the main ones for this section:

- Recreation
- Self-sufficiency
- Provision of habitats for wildlife

In addition to allocating space to different functions, it is important with a garden, as with a house, to define and structure the space, whether formally or informally, so that the different functions of each area are clear. Each garden has a given area, quota of sunlight and shelter from the wind. Different planting arrangements can provide either additional shelter, shade or sunny areas. We can also use plants to define and organise space: we can prune them into different shapes, grow them up trellises and across pergolas or we can

External use of space.

let them grow wild to produce more natural shapes. In this way we can structure our gardens with the plants themselves.

Landscaping the soil is another way we can structure the garden—it
provides a great place to try out an earth wall if you wish to experiment. If
you have a slope to contend with, you can use this to maximum effect by
terracing. Paving can also be used to structure the garden by helping to define
areas for human use. Formal shapes, such as squares, or natural shapes such
as spirals or tessalations, can also be used to structure the garden and give it
a particular character.

Different functions require connecting either formally or informally with
circulation routes. Access points, doors into and out of the garden need to
be incorporated in ways that make the most of the available space. Are there
paths that you use only for getting from A to B? Ideally such paths should
link up any paved areas so as to cut down on circulation space.

Making the most of a small garden

If you have a small garden, what can be done to make the very most of the space available? It is possible to achieve very effective results by using small scale or miniature versions of plants along with small-scale paving (such as granite sets). Another way of creating a feeling of space in a garden is through the use of light. To this end, it is useful to grow light coloured plants up vertical surfaces or, alternatively, whitewash any enclosing walls. Careful decisions need to be made as to which features or functions should be included as it is a mistake to try and achieve too many different objectives. Sitting areas may well need to be up against walls or in corners, and compost bins need to be carefully sited. Having a buried compost bin, with just a small lid showing, as some Japanese do, can be a way of dealing with this problem. If you have very little room for trees then you can either have them trained to spread flat against a wall (an espalier), or you can provide a pergola or trellis up which to grow climbers such as clematis or runner beans.

Growing your garden up walls

Plants grown up the side of a house can fulfil several different functions. They can deflect rain and keep the wall dry and in so doing add to the insulating properties of the wall and reduce possible frost damage. They can also provide an attractive covering to the facades and a safe habitat for birds and bats. And they will help to improve air quality in built-up areas.

There are several different sorts of plants that can be used for this purpose. Evergreens provide better weather protection in the winter. Ivy is the obvious example, but its growth needs to be controlled as otherwise it can do serious damage. If you provide support systems using wire or light wooden frameworks, then you can grow a far wider range of plants. For instance, privet or wistaria can also be

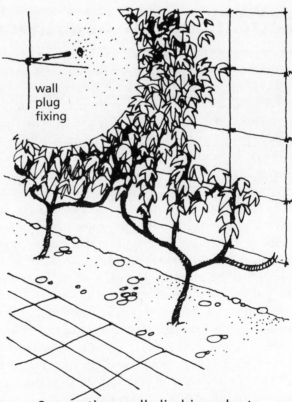

wall plug fixing

Supporting wall-climbing plants.

grown in this way and trained where you want them to grow. These climbing plants also grow to varying heights, so it is worth matching the plant to the height of the wall you want covered. When planting creepers, this should be done about one metre from the wall.

Container gardening

Window boxes are an example of container gardening which has a number of advantages. Containers can be used anywhere and, if not too heavy, can be moved around to suit the season and be protected from frost. They can also be maintained either *in situ* or at your leisure. It is important to ensure that if a container is used as a window box it is secured firmly or is heavy enough not to be blown off by the wind. Window boxes require a considerable amount of upkeep, but are a good way of providing a mini-garden for those that have no proper garden.

Containers can be used for vegetables or miniature varieties of fruit bushes. As well as window boxes there are strawberry barrels, peat-free growbags or large pots. If you have a balcony or concreted back yard, these can provide added scope.

Recreational use of outdoor space

Many people think of their garden as an outdoor area for picnicking, sunbathing or reading in warm weather. If you want to produce a space which is more intentionally a room for this purpose, you can do this very effectively by outlining the space with elements such as a trellis, columns or a specially shaped tree. In Britain we mostly want the sun to shine directly into such a space and also have it well protected from the wind. To achieve this you may have to arrange the space in such a way as to receive sun only at particular times of day or use an optional canvas sail to produce shade when needed.

As for garden furniture, it is important to decide whether you want it to stay out in all weathers, or to be brought out only when it is fine. Apart from plastic furniture, garden furniture that can withstand all weathers is either expensive or requires a lot of upkeep. Think of fold-away or stackable furniture as it is the most flexible and you can keep it in good condition out of the sun and rain. To complement this type of seating, stone or low brick walls can be used for both sitting and defining space. These will also keep the heat of the sun—garden cushions can be brought out when it is cooler.

(For many the car is still a necessity and may require parking in potential garden space. If not a garage, this space can become another enclosed room, screened from view with plants and trellises and even a pergola. It could then be used for other outdoor activities, such as an outdoor workshop, when the car is not there.)

Opportunities for self-sufficiency

Learning to be more self-sufficient is to follow one of the most important principles of ecology. The differences in the opportunities that people have to do this are dependent on the amount of time and land that is available to them. However, in the long term, all activities will have to be sustainable—at one level or another—be it household, neighbourhood, city or regional level. From an ecological point of view, the more devolved the level of self-sufficiency the better. Designing and renovating homes to achieve this end, along with the necessary local and regional changes, is one of the major tasks that we face.

What are the ways that our gardens and external areas can help this process of sustainability? One important principle is to recycle as much as possible within the confines of our gardens, and to reduce the need for ecologically expensive inputs and outputs. This is summarised below:

	inputs	outputs
Natural inputs	solar energy	vegetables, fruit, flowers and seeds etc.
	clean air (relatively)	cleaner air ($+O_2$)
	rain water	clean water (relatively)
	soil	
	plants & seeds	organic waste
	organic waste	
'Modern' inputs & outputs	tap water	polluted ground water
	fertiliser	nitrate runoff
	pesticides	pesticide pollution
	electricity	polluted air – CO_2 pollution
	petrol for machinery	from bonfires
	peat	garden rubbish to the local tip

If we look at the above table we can see in just how many ways we have introduced unecological practices. We need to reduce these as much as possible. For instance, we can replace our input of tap water with recycled 'grey' water from domestic washing. We can compost more organic waste from our homes to replace our use of peat. We can use a manual lawn mower rather than a petrol-driven or electric one. We can also find alternative ways of dealing with pests, and use plants to fertilise the soil rather than using manufactured chemicals.

The main opportunities for self-sufficiency that require allocation of space and resources in the garden are:
- Food production
- Soil conservation and composting
- Water management

Food Production

In the past most people grew their own food; now we are dependent on an unsustainable and polluting agriculture. If we do garden at all we should consider planting fruit trees and growing vegetables between the flowers. Most gardens are too small to grow all our vegetable and fruit requirements; however what we *do* grow is at least under our control, and can allow us to keep to organic principles in the process. How far is it possible to have a garden that provides a reasonable amount of food? Almost every garden has enough space for growing a small quantity of vegetables and one or more fruit trees. Herbs can of course be grown in all corners of a garden and inside the house as well.

The first decisions to be made relate to fruit trees. These can be very productive and can provide some to all of our autumn and winter fruit needs, depending on the size of our garden. If you have no fruit trees at present, it is worth planning their positioning such that shading in particular is carefully considered. The choice of fruit or nut trees (e.g. apple, pear, plum, hazelnut or walnut) is a personal one, but it is worth finding out what is likely to grow well in your local climate and soil. Fruit trees are probably the best crop to go for if you have enough space, as they require little maintenance—only yearly pruning and limited feeding.

Certain vegetables, such as broccoli, spinach and lettuce, are very productive in quite small areas. Peas and beans can be grown up hedges and garden fences. By carefully selecting the varieties, it is possible to harvest a succession of fresh vegetables and fruit throughout the year particularly with the use of a conservatory, greenhouse or cloches. If you have a small garden and wish to grow more, consider an allotment. If you are growing vegetables for the first time, start in a small way and experiment with different crops and different positions in the garden.

There is one further problem to be mastered. How can we successfully grow fruit and vegetables without modern pesticides and fertilisers on which so much food production has come to rely? Fertilisers can be replaced by compost and by an understanding of what makes a fertile soil, which are addressed in the subsection below.

As regards alternatives to pesticides, it is important to learn ecological methods of pest control. Here are some examples of what you can do:

- Encourage natural predators—spiders, hoverflies, ladybirds, and dragonflies all prey on a wide range of insect pests. Birds, toads and hedgehogs are also effective predators of slugs and snails.
- Use ashes, lime or sawdust around plants to act as barriers.
- Use aromatic plants—garlic, onions, marigolds, and tansy—these repel some insects, and the plants can also be made into natural sprays.
- Time your planting to avoid the worst seasonal infestations.

- If necessary, use weak soap solutions with small amounts of vegetable oil.

Some other ways of avoiding the use of pesticides:
- Choose only healthy plants and seeds.
- Give plants the right soil conditions, light and moisture.
- Rotate annual crops.
- Grow a diversity of vegetables and flowers.

Soil conservation and composting

Topsoil is like a living organism in itself, full of enzymes, bacteria, decaying plant materials and insects. But it needs to be kept in good condition, and is best protected from drying out by a covering of organic mulch or plant growth. Worms help in many ways, including aerating it with their holes. It can be further fed through the addition of compost.

Composting is the important link for the recycling of all organic waste from house and garden—even paper can be added in small quantities if it has been shredded. The principles of composting are simple and there are many books which advise on the best methods.

The main decisions to be made initially are what type of composter to buy or construct, and where to position this in your garden. Compost containers can easily be made from second-hand wood—an example of one is illustrated below. Alternatively, there are a large range of plastic composters

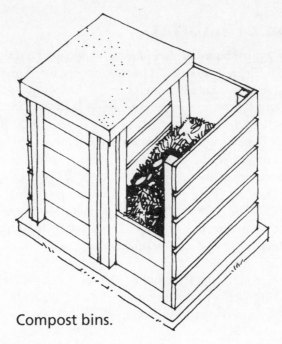

Compost bins.

on the market. Try to find a convenient place that is both accessible and not too close to the house. Shrubs can be planted around it if you wish to hide it from view.

If you have a lot of woody material that you normally burn, consider buying a shredder (ideally sharing it with neighbours). Smoky bonfires simply convert organic matter, full of potential nutrients, into a serious pollutant full of poisonous carcinogens and greenhouse gases. A shredder reduces the bulk of cuttings and prunings and produces a mulch or food for the compost.

Water management

Much depends on the part of the country you live in, as to whether water conservation is an issue for you. Even if it is not a serious problem, it is useful to be at least to some extent self-sufficient, and to have a butt of rainwater in reserve. At the other end of the spectrum there is the problem of water runoff in storm conditions that causes floods, erosion, and in the garden a loss of nutrients. Increased paving of the countryside has not helped this problem. With many gardens the water will dissipate after even the heaviest storms, but where we become aware of water run-off, it can often be diverted to a depression created for the purpose or to a soak-away.

A further step is the diversion and storage of domestic bath or shower water as illustrated on p.138. If you live in the country beyond the reach of the local sewage works, you may even want to treat your own with one of the increasing number of reedbed designs.

Provision of habitats

The habitats of a great many species of plants, animals and insects have been seriously reduced in recent years. This has occurred through the general decrease in wilderness areas and the erosion of our countryside by modern agricultural methods, urban encroachment and road building. For instance frogs and newts, which are useful predators, are in serious decline due to the reduction in ponds and wetlands. In other parts of the country, many species of plants, birds and insects that lived in hedgerows before they were grubbed up, have also been left without a suitable habitat. Even bats, which used to find a refuge in our roofs or behind timber weatherboarding, have suffered severely from a reduction in suitable loft spaces and the unnecessary universal toxic treatment of timber. In our gardens we have the opportunity to help redress the balance. If we wish, we can make conscious decisions to encourage particular species by understanding their specific needs. We can also encourage wildlife generally by avoiding the use of chemicals, by growing hedgerows and thickets or constructing a pond. And if the garden is big enough, having an area of controlled wildness can lead to interesting surprises.

As we gain more knowledge, we may want to go further and provide a more complex ecology. With insects, for instance, there are some that are pests and eat the plants we want to keep and others that are allies and are the natural predators of these pests. With a bit of study, it is possible to find ways of supporting colonies of insects that are allies by providing the right environment of plants etc. This ecological form of pest control and general line of thought can be extended to many different types of interaction between soil, moisture, plants and other organisms. It takes a different sort of knowledge than that required for understanding the labels on bottles.

Ponds

The provision of a pond is an important ecological asset; they can be incorporated into most gardens, however small. They provide habitats for many water creatures whose numbers are threatened and larger ponds can incorporate the attraction of fish and larger water plants.

Building a pond requires more than making a hole in the ground and lining it. There are numerous books which advise on their construction. To have success you need to understand how water quality can be maintained, and the needs of the species you wish to support. Alternatively you can simply provide the water and wait to see what develops! One of the valuable by-products of setting up a pond will be a better understanding of this particular ecological niche, which you will be able to pass on to others.

Boundary habitats

Nearly all our gardens have some sort of a wall, fence or hedge to mark boundaries. In the past each area of the country had its own vernacular form of barrier and in many places these are still very much a feature of the landscape—hedges, stone walls or earth mounds, or even a combination of all three. The older this feature is, the greater the number of species that become accommodated. The best boundary habitats are ones that provide the most variety. The ideal boundary habitat for wildlife would be a mound of earth with stones to keep it from falling apart, topped with a hedge, and the whole covered with plants of all types. This type of boundary is common in parts of the country such as south-west Wales and Devon. If there is a suitable opportunity, there is no reason why we should not produce our own version of an attractive vernacular boundary and provide for a wide range of species.

Birds & butterflies

Birds are attracted and supported by the provision of suitable food and nesting sites, whereas butterflies are attracted by particular plants. Tall dense hedges or thick wall creepers make good nesting sites for birds—alternatively, special birdboxes can be bought to encourage particular species. Your local ornithological or natural history society should be able to help you to choose one and to decide where to position it. If there are cats in the neighbourhood, care should be taken with the siting of any bird-table or nesting box.

Habitats for bats

There are fifteen species of bat in Britain, each of which has its own roosting requirements. Traditionally, bats roosted in trees and caves; their use of buildings has increased as woodland areas have declined. However, the advent of toxic preservatives in roofing timbers and the use of pesticides on farmland has had a devastating effect on them. They find roosts in crevices, behind tiles and weather boarding, or above soffits; and some live in hollow walls and roofspaces. They cause no damage whatsoever and can have a beneficial effect by keeping insect populations down. Since 1981 there has been protection given for bats, and it is a punishable offence intentionally to damage, destroy or obstruct access to any place that a bat uses for shelter, or to disturb a bat whilst it is occupying such a place. English Nature (or Welsh, etc.) must be consulted if building maintenance, alterations or timber preservation treatments are planned at a site which is suspected. Failure to recognise the evidence of the presence of bats is not excusable in law. It is therefore the responsibility of the householder to be aware if his or her home is used by bats.

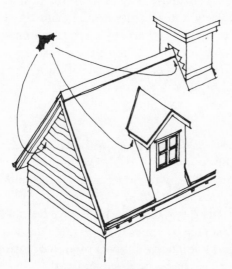

Possible bat roosts.

Most useful of all is to incorporate bat boxes or various bat access points and roosts in suitable locations—mounted high on an east- or south-facing wall. Special bat bricks are even available from builders merchants: they have special bat-sized weatherproof holes to allow bats access to roofspaces. Bat tiles can also be adapted from almost any ventilation tile by creating a small gap of around 20mm wide. It is important to create a small ductway through the roofing felt if you want to encourage them to use any unused roof space. Contact English Nature (or Welsh or Scottish) for further information and advice.

[This section on bats is largely taken from *Greener Building* (see RECOMMEDED FURTHER READING)]

Priorities for action

1. Work out carefully how you want to prioritise the aims of your garden, as regards personal leisure purposes, the self-sufficiency functions and the provision of habitats. Many of these are compatible.
2. After working out your priorities, think if you need to make improvements in defining areas of your garden, using the subsection on structuring.
3. Do you wish to plant any fruit or nut trees? These need to be chosen and planted in either spring or autumn.
4. If you plan to grow vegetables for the first time it is worth obtaining some advice either from friends or from one of the many books on the subject.
5. If you don't have one already, work on setting up a working compost system. Avoid bonfires and obtain the use of a shredder to reduce woody material.
6. Consider having your soil tested to check if it is deficient in any important nutrients and obtain advice if there is a problem. Avoid using pesticides or inorganic fertilisers.
7. If you live in a part of the country liable to drought consider ways of saving rainwater or waste water from the house (see p.138).
8. Decide which parts of your garden could become habitats for plants, birds, frogs and newts etc.

Extensions

In the **INTERNAL SPACE** and **CONVERSIONS** sections, ways of making the most of the existing space within the confines of your existing walls and roof were looked at. If you have carried out these measures and are still sure that you need to increase your accommodation without moving house, then a suitable ecological extension is your next option to consider. This section will look at the methods of doing this and show that there are sometimes more ways of extending ecologically than meet the eye.

Most extensions are built at ground level. However, any new development should encroach as little as possible on the ecological potential of the immediate environment surrounding the building or on the interests of your neighbours. Valuable garden or external space should not be lost without good reason. The last section has shown just how important these external areas can be.

If the conditions are right, then the least damaging way of extending is to create accommodation underground. This may at first sound a drastic, costly and undesirable option; however, there is an enthusiastic movement to build more underground and there are some distinct advantages. First and foremost, once the development is complete it is often not immediately apparent that any development has taken place at all. Subsoil also has natural insulating properties, both in terms of heat and sound. In addition, the shading that an above-ground extension inevitably creates is avoided.

If extending below ground is not feasible, the next option is to extend above ground. There are ways of doing this ecologically: various approaches are outlined in the second part of this section. If you keep to the guidelines stated, the alternative of building at ground level may still prove the optimum solution.

The third possibility is to build quite separately. This enters the realm of a new building, most of which goes beyond the scope of this book. However, a brief look at this problem provides an opportunity to outline a few principles. To put up a whole new building is the least ecological choice in most cases, because of the increased use of materials and additional intrusion on to virgin land. The advantage of quite new construction, however, is that you may be able to purpose-build to a much higher ecological standard than is possible to achieve by renovating an existing building. These three alternatives are now discussed in more detail.

Extending underground

Underground building usually refers to a building which is covered by earth at ground level, so that it is only from certain vantage points that any one

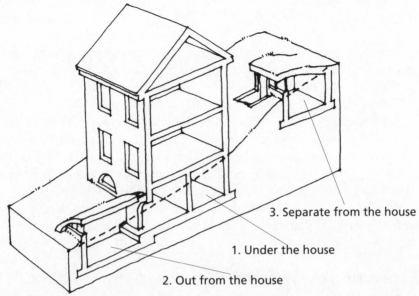

3. Separate from the house

1. Under the house

2. Out from the house

Extending underground.

would know that there was a building there. There is a popular misconception about underground buildings: to many they sound, dark, dank and thoroughly uninviting. However, if well designed, the reverse is true—underground buildings with daylight from light wells can have a particularly light and airy feel. There are, however, situations where building underground becomes much more difficult, due to unsuitable ground conditions. An experienced engineer will be able to advise in these circumstances.

There are three ways of extending underground:

- Excavating under the existing building to provide basement or cellar accommodation, so that the existing floor of the ground floor becomes the ceiling of the new accommodation
- Excavating outside, building the accommodation, then covering over with earth and re-landscaping
- Excavating into a hillside, much like tunnelling

Of these three options the most common and least problematic is the second. However we shall look at each of them.

Excavating under an existing building

Building a new basement from scratch can be an expensive operation, particularly if easy access for removing the subsoil is difficult. Additionally, all the foundations surrounding the new accommodation will have to be underpinned down to a new foundation level below the new basement floor. If the ground water level is high then the whole structure will need to be tanked. New access stairs also will need to be built, ideally under existing stairs.

To alleviate some of these problems, the best way of extending underground, if there is no existing basement or cellar, may be to extend partly under the house and partly under the ground outside that is adjacent. This will allow easier access for underpinning and earth removal.

If you have an existing cellar or basement, the situation is eased by the fact that many of the problems, such as the provision of access and foundations, will have been overcome; however underpinning to a new level will still be necessary on all the load-bearing foundation walls surrounding the new accommodation. Tanking may also be necessary or advisable.

Whichever of these options is adopted, it is advisable to obtain the best technical support you can; an engineer with experience of similar operations would be a good choice.

Excavating outside and adjacent to the building

To classify this type of underground building as an extension would mean extending from an existing basement. If not, it is virtually a separate building. We will thus assume that you have an existing basement from which to gain access to the new accommodation.

This is a much simpler operation than building under an existing building, since it is simply a matter of excavating where the extension is to be built, building the accommodation, breaking a door through to the existing basement and then covering over with earth after thoroughly waterproofing and insulating the structure. The whole area on top is then landscaped, the most important design feature being the way of bringing light from above down into the extension. This can be done either with light wells immediately above the extension, perhaps making a feature of the above ground 'roof-light', or high-level windows projecting above the ground. A third possibility is for basement-type access and windows from the garden.

Some of the ecological considerations when undertaking any kind of underground excavations are:

- Care in the removal and reinstatement of the living topsoil. This can be stored by either covering with mulch or growing a temporary covering of grass, necessary to prevent erosion and weathering of the exposed surface layer.
- Finding an appropriate home for the excavated subsoil. Ideally it should be used in the immediate vicinity for landscaping or barrier building. If it is to be trucked away, try to find a local user who has a complementary use for it.

Excavating into a hillside

If you are living on a steep slope there is the possibility of extending into the hillside. This can be achieved at whatever level seems most appropriate: at basement, ground or even first floor level, depending on the slope. One advantage of doing this is that the excavated earth can be used to terrace in front of the house.

Extending above ground

What are the alternatives when considering an above-ground extension, and how can an extension be built which will actually enhance the existing buildings' ecological performance? Many extensions that are made without regard to ecological principles project and intrude where they should not, appropriating sunlight and more green space than necessary. An ecological design is essentially compact and attentive to the needs of the site.

Here are some of the principles to consider applying:
- An extension to the north should be used to improve the insulation of the north-facing walls and windows.
- If the roof slopes down towards the north then an extension in that direction should ideally continue the line of the roof down towards the ground in order to reduce the amount of shading at the side of the house. Furthermore, this reduction in shading can be further reduced if it is possible to raise the ground around the house at the back. This is referred to as berming, and is illustrated below.

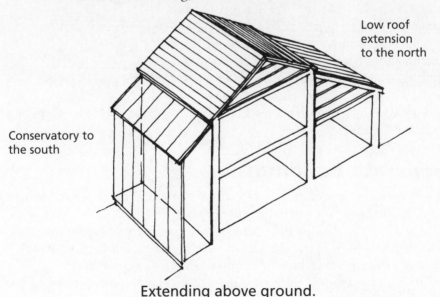

Low roof
extension
to the north

Conservatory to
the south

Extending above ground.

- An extension to the south should use this aspect to make the best possible use of passive solar gain, with the type of glazing used for a conservatory or sunroom. The extension can then be designed to ease the heating load on the building rather than to increase it. (Further information on ways of using this passive solar energy are discussed in the last chapter of **ENERGY**.)
- An extension can be created by glazing over a space between walls or buildings. This has the effect of consolidating the overall mass of the building.

All extensions should take account of the prevailing wind and shield any entrances from this direction. Also it should be clear, before undertaking any building work, exactly how the shadow of any new extension will affect any neighbouring property or important external feature.

It should not be forgotten that there are limited but effective ways of extending from the first or second floors without affecting the external ground level. It is possible to build a light-weight cantilevered structure, not unlike a bay window, at first floor level, or an enclosed balcony. This feature is called an oriel window and was used more commonly in the past. It can have all the advantages of a sunroom or conservatory. The concept is illustrated on the right.

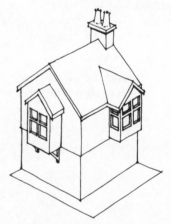

Dormer windows are also a form of extension and can be used to enlarge and give more headroom to an attic space. These too can be designed as sunrooms, conservatories or greenhouses as required. Finally, it is worth thinking carefully how you can alter your home without spoiling its character too much.

Oriel windows.

Is it possible to incorporate some of the old vernacular methods of building from the locality?

Separate buildings

If you decide you need an entirely separate building, the problem changes dramatically from one where the site, the style, the size and the orientation have all been decided for you, to one where you are starting from scratch. The problem is complex, because of this increased number of variables. However the same basic principles apply as with extensions.

First and foremost is respect for the site you are intending to build on.

This requires looking at carefully, to understand its true ecological potential. Is the soil rich and fertile, or is it barren and polluted? These are extremes, but the message is clear: we should always try to build where it is difficult to cultivate. Our tendency is to do the opposite and much prime quality land has been built over. At present our use of vast amounts of synthetic fertilisers and pesticides is masking the true situation and gives us the illusion that we can grow whatever we like in whatever quantities we like. This practice is, however, unsustainable.

There are many other criteria that we can apply to this respect for the site, such as the spoiling of views, the destruction of historically important sites, the destruction of trees, the local importance that may be attached to the site as a special area for wildlife, and so on. Apart from the building itself, it is important to work out how much land will be required for access roads and paths.

In some cases these considerations will lead you to build largely or wholly underground to reduce the impact of the building to its minimum. Whatever the outcome of your deliberations, finding the right place ecologically should be the starting point.

In terms of its construction, there are many books on new energy-efficient houses which should lead to a building that is super-insulated and uses the sun's energy to maximum advantage. With the application of existing knowledge, the end result should be a building that uses a fraction of the energy that the average building uses now.

Finally, a new building should use materials that cause as little damage as possible both to the environment, and to your health as a result of living in it. Both these aspects are investigated in Parts Three and Four.

Priorities for action

1. Use all your existing internal space first, and then make any conversions that are possible within the basic framework of the house. (Refer to the **INTERNAL SPACE** and **CONVERSIONS** chapters.)
2. If you still want to extend, analyse carefully whether this cannot be accomplished underground by looking at the various options outlined above.
3. If you are extending above ground, plan an extension that will add to the ecological functioning of the building as a whole.
4. If you are building to the north, north-east or north-west, design a structure which reduces the shaded area, is highly insulated and uses smaller windows.

(continued over)

(Priorities for action continued)

5. Extensions to the south, south-west and south-east, if not overshadowed, should be built so as to benefit from the full use of passive solar energy to provide living space or growing space that requires no additional heating.(See further information on this aspect in the SOLAR ENERGY chapter of the next Part.)

6. If, after careful consideration, the only way of extending is a new, separate building, then respect for the site is the most critical ecological consideration. Ecological approaches to energy efficiency, health and materials, as outlined in this book, can all be applied in a more complete way than is possible with the renovation of an existing building and full advantage should be taken of the opportunity.

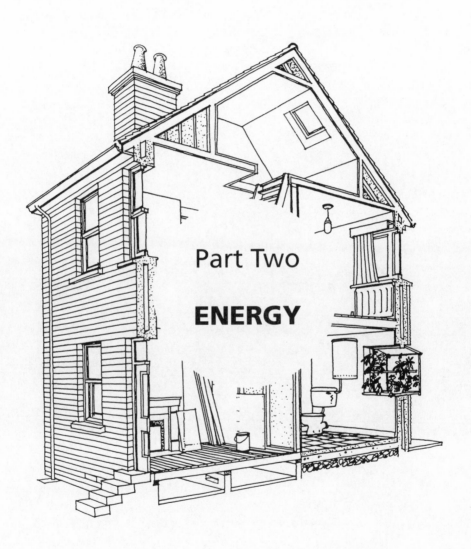

Part Two

ENERGY

Part Two

ENERGY

WE LIVE IN A WORLD WHERE energy has never been so cheap and easy to use. This has led us to waste it on a massive scale, with results that are becoming clearer almost month by month. As the real ecological costs are not included in the prices we pay, we are being sold this energy at an enormous discount. The costs of global warming, resource depletion and acid rain are impossible to calculate, as we simply do not know what their final effects will be.

By general consensus, global warming is one of the most dangerous instabilities that we are introducing into our ecosystem. So what is the relationship between global warming and the way we use energy in our homes?

Global warming

There are four main greenhouse gases that cause global warming: carbon dioxide (CO_2), methane (CH_4), the chlorofluorocarbons (CFCs) and nitrous oxide (N_2O). Carbon dioxide's warming contribution is about 50%, whereas the contributions of methane, the CFCs and nitrous oxide are approximately 18%, 14% and 6% respectively. The surprisingly large contribution of these latter gases is due to the fact that methane is 25 times more effective than CO_2 as a greenhouse gas, CFCs up to 10,000 times and nitrous oxide 150 times. Nevertheless carbon dioxide is the most important, all the more so for us because it is integrally bound up with the carbon cycle and our use of fossil fuel. The amount of the sun's energy falling on the earth has increased about 25% since the beginning of life; carbon dioxide has played an important role in moderating this increase, through its gradual absorption by plant forms, mainly in the sea. We are now in danger of destroying this delicate natural balance by exhausting ever increasing quantities into the atmosphere. The results could be catastrophic.

During the energy crisis in the 1970s, the concern was with our diminishing reserves of gas, oil and coal; the concern now is to reduce the quantity of CO_2 that reaches the atmosphere—we may never be able to use our reserves of fossil fuels to the full!

What can we do to halt this trend? Our daily patterns of consumption in western 'civilisation' provide much of the answer. The decisions we make to boil a kettle for a cup of tea, take a bath, turn up the central heating or drive to the supermarket all contribute. We spend the major part of our time in our homes and much of the rest we spend in our cars and transport systems or working in factories and offices, often in order to acquire and consume even more energy-consuming goods.

Estimated CO_2 emissions by end use (UK)

Households	28%
Industry	26%
Service sector	14%
Road transport	24%
Other transport	7%
Agriculture	1%

Source: BRE 1996

The above table is interesting in that it points to well over a quarter of the CO_2 produced being caused directly or indirectly by the use (and misuse) of energy in our homes. It is clear that the way we use energy is extremely inefficient. If all the measures outlined in this chapter were carried out the amount of CO_2 produced could be reduced by 75%. If we were also to change our lifestyle and use renewable sources we could reduce this still further, perhaps to as low as 5% of our original CO_2 emissions.

The house as an energy system

It is useful to think of the house as a total system when it comes to the way energy behaves. Firstly we can look at the overall amount and form of the energy that is entering the house. Then we can see the ways this energy is being used, and finally we see how the resultant heat is lost to the outside:

Input (sources of energy)	Use	Output (losses)
Fossil fuel: Gas	Space heating	Draughts, ventilation
Oil	Water heating	Hot flue gases
Coal	Lighting	Heat loss through:
Electricity	Appliances: Cooking	Walls & windows
Biomass: Wood	Washing	Roof
Solar energy: Passive	Human metabolism	Floor
Active		Waste warm water
Food		

This diagram helps us to understand what is happening, and to work out the ways we can reduce our carbon dioxide emissions to a minimum. The important point to notice is that a saving can be made in CO_2 emissions at every stage in the system.

This can be done by:

- Choosing the sources of energy emitting the least amount of CO_2
- Finding the most efficient ways or using the energy in our heating systems, lighting and appliances
- Conserving the heat in our homes through draught-proofing, ventilation control and, most important of all, insulation

These different aspects are all looked at in this chapter, including the storage of heat energy, either in the building fabric or in hot water—both of these are seen as increasingly important once we are using insulation to full effect. A further aspect of the energy equation is that used in the manufacture of building materials. This is addressed in the chapters on **MATERIALS**.

Improving the overall energy efficiency of our houses is no longer controversial. It is only a question of the resources that can be directed to the task and how far to go with each measure. This is a case where investing pays off for everyone!

Sources of energy

Which sources of energy are least harmful to the ecology of the planet? It is the purpose of this chapter to establish an order of priority when choosing a form of energy for a particular function or appliance. The most important of the criteria to be applied will be the relative amounts of CO_2 produced by different energy sources. Other environmental factors mainly involve other forms of air pollution such as sulphur dioxide, one of the causes of acid rain. This chapter will start with an analysis of fossil fuels, continue with a separate look at electricity, and then make comparisons between these different energy sources. It will end with a look at renewable energy.

The fossil fuels

There are three basic forms of fossil fuel:
- Solid: coal and its products
- Liquid: oil and its products
- Gas: natural gas and liquified petroleum gas (LPG)

All fossil fuels are organic in origin and thus contain carbon to a greater or lesser degree. This means that there is no way of avoiding emitting CO_2 when we burn fossil fuels. However, there are considerable differences in the proportion of carbon in each of these different fuels and thus in the amount of CO_2 emitted. When we compare the burning of coal with the burning of oil or gas we arrive at a very interesting and relevant comparison. (It is assumed that coal is pure carbon, oil is represented by approximately one carbon atom to two hydrogen, and methane is CH_4, a carbon-hydrogen ratio of 1 to 4):
- Coal/coke (carbon): $C + O_2 = CO_2$
- Heating oil: $C_2H_4 + 3O_2 = 2CO_2 + 2H_2O$
- Gas (methane): $CH_4 + 2O_2 = CO_2 + 2H_2O$

If we compare the above chemical reactions we can see that the main product of combustion of coal is CO_2, whereas with both oil and gas, water (which is ecologically neutral) is produced as well, indicating a less polluting burn. In fact, as a rule of thumb, the higher the proportion of hydrogen (a high energy burner) in a fuel, the less harmful it is. The ultimate ecological fuel is hydrogen (in which the only product of combustion is water); there is a growing lobby for its development linked to solar power. We thus have a range of fuels with carbon (coal) at one end and hydrogen at the other.

Solid fuels (coal and coke)

The use of coal presents us with another problem. Besides burning with the highest proportion of CO_2, coal burning can be polluting in other ways. The constituents of coal vary enormously from anthracite, which is around 94% pure carbon, to brown coal or lignite, which contains many other compounds including sulphur. Most of the coal in Britain lies somewhere between these two extremes. When burnt, not only CO_2 but also sulphur dioxide is produced; this combines with water to form sulphurous acid—the main cause of both smog and acid rain. In Britain, the clean air acts of the 1950s and 1960s, which introduced smokeless fuel zones, reduced the sulphur pollution in the big cities; but this pollution was transferred to the upper levels of the troposphere (the lowest layer of the atmosphere up to 18 kilometres) via the tall chimneys of coal-fired power stations and coking plants. From being a local problem of smog in large cities, it became the regional/global problem of acid rain.

Fewer and fewer people are now burning solid fuel in their houses, for reasons of dirt, smell, pollution, inefficiency, time and cost. This trend is likely to continue, and from an ecological point of view should not be discouraged. Burning coal in an open hearth, where 90% of the heat goes up the chimney, belongs to a bygone age. If, however, you are heating your house with one of the latest high-efficiency solid fuel boilers, you are doing less damage to the environment than if you used electricity for the same purpose.

Liquid fuel (oil and its derivatives)

In Britain, oil had its heyday as fuel for heating during the 50's and 60's; a combination of the discovery of North Sea gas and then the '73 energy crisis put an end to its popularity. If we look at the chemical reaction in the last section, we can see that when oil is burnt approximately one molecule of water is produced for every carbon dioxide molecule. This points to oil being a better fuel to burn than coal in terms of carbon dioxide emissions. On average the sulphur content is also considerably less. This makes oil a more benign fuel than coal to burn in terms of acid rain and other pollutants; but we need also to take into account the other environmental costs of extraction and transportation. The main environmental cost is that of marine pollution, which is a growing concern.

Liquefied petroleum gas (propane and butane)

LPG is a product of oil; it exists in two forms, propane (C_3H_8) and butane (C_4H_{10}). These are both members of the paraffin series of which methane is the first member and heating oil is about the sixteenth. Both have a carbon content closer to oil than to natural gas. Energy costs for its distribution are higher than for natural gas, and similar comments can be made about its polluting effects on the sea as with heating oil. The most interesting recent development in the use of LPGs is as a substitute for the chlorofluorocarbons (CFCs) in refrigerators.

Natural gas (methane)

Methane is the cleanest of all the fossil fuels to burn as there are two molecules of water produced for every one of carbon dioxide. It has very few impurities and its cost in terms of extraction and distribution is only 7%, which makes it the most efficiently distributed of all the fossil fuels. This is because of the huge network of gas mains that crosses Britain. The efficiency of appliances that use natural gas can be of a very high order (as will be explained in the heating chapter—condensing gas boilers can extract up to 95% of the available fuel heat in peak conditions). However there are also gas appliances with very low efficiencies, such as the mock coal fires with open fireplaces, where most of the heat goes straight up the chimney.

Natural gas is therefore the most ecological choice amongst fossil fuels, but we should treat it very sparingly as we may have little more than 30 years accessible reserves. However it could become a renewable source for the future: in China methane gas is commonly manufactured from agricultural biomass. We could do the same.

The other side of the coin is that methane is also an increasingly serious greenhouse gas: a thousand million tons is released into the atmosphere every year from rubbish tips and agriculture, contributing nearly 20% of global warming. If a higher proportion of this gas could only be collected and used before escaping into the atmosphere we would be solving two problems at once.

Electricity

Electricity in cables is a very convenient way of transporting energy. It is also fundamental to the quality of modern life: it can be produced in so many ways, and has so many uses for which there is no substitute. It can be produced by burning fossil fuels in generators, by using rainwater and the force of gravity (hydroelectric power), by wind turbines, and directly from the sun in photovoltaic cells. It travels almost instantaneously down power lines above or below ground.

Once it arrives at the place it is required, it can either be transformed into another type of energy, such as light or motive power, or used directly in computers or telecommunications. Heat is nearly always the end product of using electricity but there is increasing—and misguided—encouragement for householders to use electricity for central heating and water heating. The charts below show just how short-sighted this is.

The percentages of electricity produced from different fuel sources in the UK (1993) are approximately as follows:

Coal	40%
Oil	30%
Gas	20%
Nuclear	8%
Hydro	2%

It is possible to make a comparison between the different sources of fuel and the amount of CO_2 produced:

Fuel	Kg of CO_2 emitted per gigajoule delivered (approx)
Natural Gas	53.9
Domestic Oil	79
Domestic Coal	81.4
Electricity	141.6

Source: BRE Communication 1996

From these two charts it is possible to see just how polluting electricity is. The second chart shows that electricity produces almost three times the CO_2 pollution that natural gas does for the same amount of heat; almost twice as much as oil and coal. If we were to compare electricity generation purely from coal we would find an even worse picture: it is only because there is considerable electricity generation from oil, gas and nuclear power that the picture looks better. Besides the pollution from coal-fired power stations, much of the energy produced is thrown away in cooling towers. In addition there are energy losses in the national grid. It is for these reasons that the efficiency of electricity in terms of the carbon dioxide produced per giga-joule of power is so very low.

What then would be a positive ecological strategy for the electricity industry? The following measures would all contribute:

- Drastically reduce the use of electricity for heating except where there is absolutely no ecological alternative.
- Introduce the latest energy-efficient technology in all areas of high electrical usage.

The above savings in energy would be more than enough to enable the phasing out of inefficient existing coal-powered generating stations, and to bring the remaining up to a high standard of energy efficiency and of pollution control.

Also, waste heat should be used for district heating where possible.
- There should be massive investment in wind energy generation with a plan to replace electricity generation from carbon dioxide producing stations.

All the above measures are technically feasible today: it requires only the political will. The carrying out of these measures would mean that electricity users could then use electricity with a clear conscience. Paying the real price for it would also encourage care in its use and the development of more efficient appliances.

Electricity, as it is presently generated, is a non-renewable energy source and the most polluting form of power in terms of the production of carbon dioxide and acid rain. It is only now beginning to be realised that the cost of conserving energy is far less than the real cost of producing it, let alone the cost of building new power stations.

Renewable sources of energy

What are the renewable sources of energy that are available to the householder? The most obvious is that from the sun—solar energy. This is so important a source that it has been given a chapter of its own (see p.102). The next most favoured form of renewable energy is that of organic material in one form or another, known as biomass. The burning of wood is the most common example of the use of biomass. All forms of renewable energy, except geothermal energy, are indirect forms of solar energy. Biomass stores the energy from the sun through photosynthesis. There are other ways of using biomass energy besides burning; these are mentioned below. Lastly, wind or water power from from small-scale wind turbines or water turbines is available to the few people who have a suitable location and enough money for investment in the technology.

Biomass

Biomass is a shorthand term for organic material which can be used to create energy. Traditionally it has been the main method of heating and lighting, through the use of wood and inflammable organic matter. Now it is seen as an increasingly important way of gaining energy from organic waste without necessarily burning it. There are a number ways of converting biomass into energy:
- Direct combustion—of dry organic matter.
- Pyrolysis—heating organic compounds, either in the absence of air or in the presence of oxygen, water vapour or hydrogen, to produce different gases such as methane, carbon monoxide or hydrogen.
- Anaerobic digestion—biogas (methane) is given off when wet sewage

sludge, animal dung or green plants are allowed to decompose in a sealed tank under anaerobic (oxygen-free) conditions.

- Fermentation—alcohol is produced by the fermentation of sugar solutions (sugars can be obtained from many different processes and products, even from cellulose). After fermentation the solution is distilled to extract the alcohol.

A combination of these methods combined could provide a considerable proportion of our energy requirements in the future, and could lead to the substitution of natural gas by renewable biogas. It is doubtful whether any of these methods, except the wood-burning stove, will be applicable at a domestic level, but on many farms there is a significant potential for using agricultural waste in methane digesters to meet most, if not all, of the farm's energy requirements.

Wood burning

Wood, if used in a sustainable manner, should not contribute to global warming. However if we look closely at our use of timber on a global scale we see that we are cutting down many times the quantity of trees that we are planting. In Britain, we are net importers of large quantities of timber. In these circumstances it is difficult to see whether wood can be seen as a renewable energy source. However there is so much waste wood at present available that burning it efficiently is a useful service. So if you are using a wood stove, endeavour to feed it with scrap wood or wood from renewable coppicing.

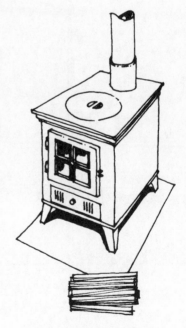

Wood smoke can be very polluting: the worldwide use of wood for burning contributes millions of tons of pollutants. These pollutants can be considerably reduced by very careful control of your wood stove to ensure as clean a burn as possible. There are now high-efficiency stoves available; however, they still require careful tending and a good store of well-dried wood.

Wood-burning stove.

Wind and water

The possibility of generating power from wind and water on your own land is available to very few households. However, there is often great interest in these sources, so they have been included mainly for information. It may be that in the future a suitable windmill will be designed for use on houses, but that moment is not yet here. Historically, wind and water have been important sources of energy. In the USA over 8 million mechanical windmills have been installed since the 1860s: in the 1920s and 1930s, before subsidised rural electric power lines, many of these were wind generators, each delivering 200-300 watts, and providing all the electrical needs of outlying farms at the time. In the 17th, 18th and early 19th centuries in Britain there were thousands of windmills and watermills, which formed the main source of motive power.

Wind energy is almost certainly the most effective way of generating

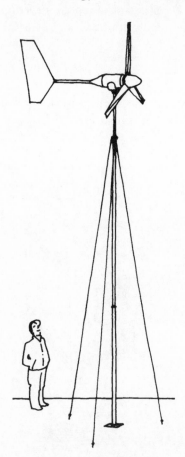

Wind generator.

renewable energy in Britain, particularly as in mid-winter, just when our energy requirements are at their greatest, the amount of energy that can be harvested from the sun is reduced to almost zero. On a domestic scale, the main requirement for a successful wind turbine is having a good location to site the mast. Unfortunately, attaching one to the top of your house, unless you have a specially designed chimney stack, is not recommended because of the buffeting it will receive and the vibrations it will generate, quite apart from your neighbours' likely concerns. At present, wind turbines are only suitable for housing with enough land and a suitable location: they need to be positioned as high as possible and away from any obstacles such as buildings and trees that would interfere with their efficiency. It is clear that not many sites in built up areas would be suitable.

Proper advice about installing a wind generator is beyond the scope of this book. However, the following points may be helpful:

• A wind turbine should almost certainly not be attached to your house as the vibrations and noise would be a problem, unless the structure has been specially designed for it.

• Nothing should interfere with maximum winds. Any tower should extend so that the turbine reaches six metres above any obstruction for at least 150 metres all around. This is a severely limiting condition and will preclude the use of wind power for the majority of locations.

- Any wind plant should ideally be located within 30 metres of the house to reduce transmission losses.
- Towers must be strongly built.
- The wind should be tested for at least three months beforehand to check feasibility.

It is also necessary to do a considerable amount of homework (in the absence of professional advice) if you wish to set up your own wind turbine.

Opportunities for using a water turbine are even more remote for the average householder as a hillside with a decent head of water, either a small amount over a long fall for a turbine or a large amount over a short fall for a water-wheel, is required (see REFERENCES in the back of the book).

Priorities for action

1. When deciding on a choice of fossil fuel, choose one which creates the least CO_2 per unit of energy delivered. For most of us this will be natural gas or bottled gas.
2. Reduce your reliance on electricity to the minimum. On average, electricity releases up to 4 times the amount of carbon dioxide per unit of energy delivered, when compared with natural gas. Use it only where it is the most energy-efficient choice, and in particular avoid using it for heating if at all possible.
3. If you are using solid fossil fuel, work towards phasing out its use and finding a more ecological solution.
4. If you use wood for heating, try as far as possible to use waste wood that would otherwise be thrown away. Alternatively, use a source from a properly managed coppice. Also be prepared to adjust the stove to achieve as clean a burn as possible.
5. Remember that the cheapest source of energy is conservation, so before investing in any more expensive form of energy collection or generation, such as solar panels or even a wind-turbine, properly draught-proof and insulate to the maximum degree possible (see the following chapters).

Draught-proofing and ventilating

Draught-proofing is one of the first energy conserving measures to undertake, if you have not already done so. It is one of the easiest, cheapest and most effective improvements of all. If, however, it is spring or summer, concentrate on improving your ventilation strategy first. Ventilation needs to be attended to either before or in parallel with your draught-proofing since it is necessary to ensure that you have adequate *controlled* ventilation where you need it.

This chapter gives a more in-depth look at the causes of draughts than is usual, in order for you to have a better idea what you may be up against. The average home in Britain has a total of about two square feet of air leaks—that's when you add up all the gaps around doors, windows, service entry points and the possible hidden airways up through the middle of your house that you may not know about. These unwanted draughts and leaks are known technically as infiltration, and will vary according to many factors, such as how well you secure your windows or doors; but at whatever level they occur

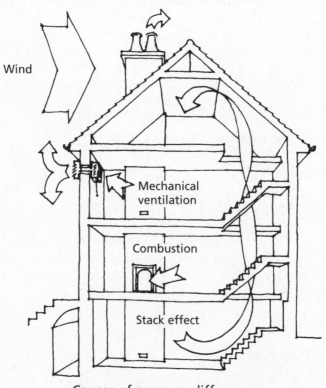

Wind

Mechanical ventilation

Combustion

Stack effect

Causes of pressure differences.

they are essentially continuous every winter, day and night. Just think how concerned we are about a leaking tap—and yet heat leaks are far more costly. Lack of draught-proofing can account for anything from 10% to 30% of your energy expenditure. (If we were to add up the national total it would amount to thousands of tons of CO_2 —millions of pounds all spent unnecessarily in heating the sky!)

In winter the benefits of draught-proofing are immediate. You no longer have uncomfortably cold draughts from the outside, and it will immediately feel warmer. You feel freer to move around and even use places where you would never usually sit because of draughts. You will find that if you use the right methods, your windows will stop rattling and you will also find your home quieter and less dusty.

The driving forces of air flow

Air will only pass from the outside to the inside, or from the inside to the outside of a building, if there is both a passage for it to pass through *and* a pressure difference between the two sides. There are four main causes of pressure differences:

- Wind
- The stack effect
- Combustion
- Mechanical extract fans

Wind

Wind can cause pressure and suction on different parts of the house: these pressures and suctions depend on both the wind speed and the configuration of the buildings and trees etc. around them. The higher the wind speed, the greater the pressure differences.

The stack effect

The stack effect occurs because warmer air is lighter than colder air. Thus in winter and at night, whenever there is a big temperature difference between the inside and the outside, there is higher pressure at the top of the house, forcing air out, and a correspondingly lower pressure at the bottom, sucking air in. These pressure differences are increased still further with each increase in the number of storeys. Research has shown that we can make the greatest difference by concentrating our attention on our basements and attics.

Combustion

Combustion in heating is another of the forces that can contribute to low pressure within the house. The worst problems occur with open hearth fires, or with gas fires or wood-burning stoves fitted to existing fireplaces. In these cases the flue acts to produce a powerful suction that draws air up the chimney—precisely what the chimney was designed to do in the first place!

Ventilating fans

Ventilating fans, particularly those expelling air from kitchens and bathrooms, are another cause of low pressure. These fans need to be balanced with incoming air and it is worth planning how this can be done in the most controlled way. There are two further points relating to air flow which are useful to bear in mind:

- *Air flow will always take the path of least resistance.* The greater the size of the opening, the more air will pass through. This only highlights the importance of ensuring that the largest voids are attended to first and that we discover where they are. Especially in older houses, air flow can occur through vertical channels up through the house interior. This is particularly true of wood-framed structures, but it can also occur in internal stud partitioning that connects top and bottom with the floor cavities. The same thing can apply to staircases and vertical service runs that are boxed in and allow large volumes of air to flow from floor to ceiling.

- *Incoming cold air draughts are balanced by outgoing warm air leaks.* This is another

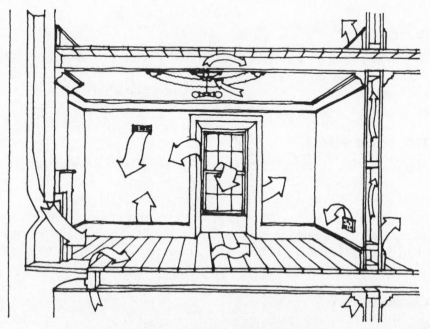

Paths of least resistance.

obvious fact that we are often not aware of—simply because we don't feel outgoing warm air leaks as we do incoming cold draughts. In two- or three-storey buildings it is often noticeable that there are draughts at ground floor level, but these are not so apparent in the upper storeys or attic unless it is windy. We never go outside and up a ladder to the second floor windows on a cold day to check for hot air leaks!

Assessing the extent of infiltration

You probably have a very good idea as to how draughty your house is. When there's a storm, do the lampshades wave in the wind ? How easy it is to find and fix the leaks depends to a large extent on the type of construction. If it is a standard brick-built house with plaster covered brick partition (internal) walls, then your problem is probably confined largely to the openings in the basic structure—the windows, doors and ground floors. However, if your house is of timber construction or built with a composite material, there are likely to be more passageways for the air to find its way about.

Carrying out draught-proofing

There are many books which cover the materials and techniques of draught-proofing well. Here are some practical suggestions of what to look out for in different parts of the house:

- Attic or roof space: look for any passageways leading up through internal partitions and surrounding services. Properly seal the door or hatch leading to the attic.
- Ground floor/basement: there should be adequate ventilation under the floorboards and cross-ventilation in any basement, to keep these spaces dry. The ventilation spaces should, however, be sealed off, similar to the treatment of the attic space above. The floor and skirting boards should all be draught-proofed along with any door leading to a basement.
- Windows and doors: start with doors as these are likely to prove the worst offenders. The most complex draught-proofing occurs in a sliding sash window: an illustration showing one way of draught-proofing such a window is shown shown on the next page.
- Fireplaces and chimneys: seal fireplaces but also consider installing a ventilator here as a means of introducing controlled ventilation. Chimneypots should also be capped with ventilated caps, ensuring that the inside of the chimney remains dry.
- Draught-lobbies and door closers: in houses where there is the possibility of wind blowing through when both front and back doors are open, consider introducing a draught-lobby or putting door closers on appropriate doors.

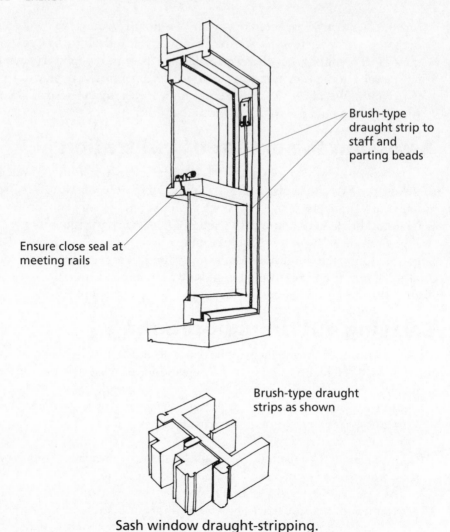

Brush-type
draught strip to
staff and
parting beads

Ensure close seal at
meeting rails

Brush-type draught
strips as shown

Sash window draught-stripping.

At the same time as draught-proofing, you need to consider an overall
approach to controlled ventilation.

Ventilation

If we sealed our home completely, we should only get fresh air inside when
we opened a door or window. We need ventilation to exhaust unwanted
smells, water vapour and pollution, and replace them with fresh air, but we
also need to control ventilation so that we can turn it on and off as we need
it, and direct it where it is necessary. In this way we can prevent the waste
of heat in the way that water is wasted through dripping taps. We then only
lose the heat in air allowed to escape for ventilation purposes.

Conventional wisdom has established that an average-sized room requires at least one air change per hour when occupied. However this varies and is dependent on such factors as the number of occupants and the number and nature of the sources of pollution. Traditionally, ventilation was achieved with the use of air bricks and infiltration; however, as our energy conserving becomes more sophisticated, we need to develop a correspondingly more sophisticated ventilation strategy. Before listing the possible measures in a strategy, we shall look at ventilation for combustion and heat exchangers.

Permanent ventilation for combustion

It is a statutory condition that heating appliances which require air from inside a room for safe operation should have a permanent ventilator. The danger is that the fuel does not burn efficiently without sufficient oxygen; if toxic products of combustion are not exhausted, they can build up in a room and possibly prove fatal. In old houses the original ventilators are often papered over and it is obviously important that either they are unblocked or an alternative route is found for the incoming combustion air. One way of providing this alternative route is via a purpose-built duct delivering air directly to the appliance. Many modern appliances overcome this problem by having a balanced flue which draws air from the outside and expels it through the same fitting.

Heat exchangers

Is there any way we can save the heat lost through controlled ventilation? Heat exchangers are designed to do just this. They are a relatively new method of recovering the heat from warm air before it is exhausted to the outside, and are being used increasingly as part of an overall strategy for ventilation and energy conservation. The principle is simple: the outgoing air is extracted through a matrix of hollow tubes and fins which warm the incoming air contained within them. In larger systems, warm air is collected via ducts from various places around the house, such as bathrooms and kitchens, and the warmed fresh air is delivered to the living rooms. The heat exchanger can be placed anywhere in the house but the roof space is the usual location. Expert advice is essential if you are thinking of installing a heat exchanger.

Your ventilation strategy

Once you have identified individual problems in each room of the house, such as a heater requiring combustion ventilation or a room with too much humidity, it is necessary to draw up a ventilation strategy. Perhaps the most important decision you should make at the very beginning is whether to install a heat exchanger with ducts to various parts of your home. If you decide this then the problem is more or less solved in one go. This should be the most energy-efficient option. If not, consider all the measures below and try to balance the air flow in each room of the house so that you have an inflow and an outflow. If this seems complicated, persevere and find ways of simplifying the problem in your mind: for instance, if you fit controllable trickle ventilators to all your windows, leave gaps round the internal doors and install extractor fans in the bathroom and kitchen, this would be sufficient. You will of course always have the option of simply opening windows as required. It is up to you how sophisticated a system you devise. Remember that in a tall house in very cold or very windy weather, whatever system you have will need to be closed right down as the pressure differences will force air through much smaller openings. Whatever you decide, it is important to develop a ventilation strategy that fits your home the way you use it. These are the possibilities for you to consider:

- Decide whether to install a heat exchange system.
- Fit controllable trickle ventilators in each room to obtain cross ventilation (the ease with which these can be fitted to existing windows varies with the type of window).

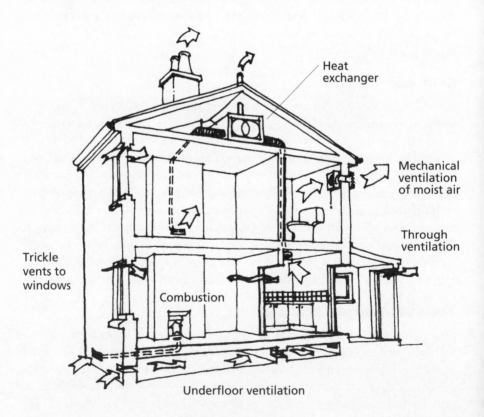

Ventilation strategies.

- Install mechanical extractor fans in kitchen and bathroom, controlled by a timer or humidistat (moisture control switch).
- Install permanent ventilation for combustion appliances which rely on a supply of air from inside.
- Use defunct chimneys as channels for ventilation or ducting. This may be particularly useful if it is difficult to fit ventilators to the windows. Consider also using your chimneys to recirculate warm air to upper storeys or vice versa.
- Install air-cleaning measures: either mechanical or biological (the use of plants—see p.157), ionisers (see air quality section p.127) or filters. If the main problem is humidity then consider using a dehumidifier and if lack of humidity then consider a misting humidifier or again the use of plants.
- Opening and closing windows as necessary: if external doors are constantly being used, this may provide sufficient ventilation for much of the day.

Priorities for action

1. Take each room in turn and look at both its draught-proofing and ventilation requirements to obtain an overall appreciation.
2. If it is autumn or winter, start draught-proofing right away using the checklist on p.51 as guide. You can always use openable windows as an interim means of ventilation.
3. Work out your ventilation strategy by following the above subsection. Don't worry if it is not clear immediately—just focus on solving one problem at a time.
4. Decide whether to go for a heat exchange system (see **INFORMATION ON PRODUCTS AND SERVICES**).
5. Ensure that ventilation is provided for any combustion requiring a supply of air from inside the house.
6. Use mechanical ventilation to vent humid air from your kitchen and bathroom controlled by either a timer or a humidistat.
7. In your draught-proofing, if you live in a two or more storey house, it has been found that the most effective places to start are at both the top and bottom of the house first, to reduce immediately the gaps and cracks where the pressures are greatest. Draught-proof the rooms you keep the warmest.
8. Work out carefully where you may have hidden air leaks: you will know them by the accumulated dust. Also go for the biggest holes first, like the undamped chimney or hole in the ceiling that you never got fixed.
9. External doors are likely to be a worse fit than the windows, so check these before starting on the windows.

Insulating

Insulating is the most important of all energy-conserving measures, because it is with the use of insulation that we can have the greatest impact on our energy expenditure: for the average house we can reduce the amount of heat lost through the fabric of the the house by at least half. If we also undertake all the other energy-conserving measures in this chapter, then on average the annual CO_2 emissions resulting from domestic energy use can be cut to a quarter of what they are at present. However this can only be achieved by taking the measures suggested in each section seriously—this is particularly true of this insulation chapter. The problems that arise in installing adequate amounts of insulation in existing buildings vary enormously. The question is no longer whether to insulate or even how much to insulate: we simply need to find the most effective way of insulating each element of the home to the highest standard that can be afforded. The purpose of this chapter is to give the background necessary to approach this task with confidence.

There are many positive benefits to come from insulating your home, as well as a few pitfalls to watch out for. Besides the environmental benefit of reduced carbon dioxide emissions, there will be increased comfort, and a more even temperature around your home. Also, your heating system can be scaled down and radiators can be more freely placed anywhere in the room and, in some cases, dispensed with altogether. Your house may also feel quieter.

The main pitfall of insulating is that of creating cold bridges—where the original construction remains uninsulated and cold whereas the newly insulated surroundings are now warm. This can sometimes result in condensation. However there are solutions to most problems, and that is just as true for coldbridging as for anything else.

Principles of insulation

There are very few principles to understand with insulation, and there is one concept that is worth bearing in mind:

TOTAL HEAT PRODUCED IN A BUILDING = TOTAL HEAT LOST TO THE OUTSIDE

All that insulation does is to slow down the rate of heat loss, so that less heat is required to maintain the same internal temperature. When we insulate our houses to a high standard, we not only conserve the heat from our heating appliances but also all the other sources of heat, such as from cooking and lighting, play a more important role.

The value of external insulation

If we were able to throw a thick overcoat around a house, it is easy to imagine that this would be a most effective way of keeping it warm. This is nearer to the truth than you might think. Theoretically, the best position for insulation is on the exterior of a building: it has the effect of keeping the whole fabric of the building warm and dry and raising the thermal capacity of the building so that uneven fluctuations of temperature are reduced. Of course it is necessary to provide a weatherproofing layer to keep out the rain, but generally it is worth bearing this principle in mind and putting it into practice wherever there is the opportunity.

Condensation

This occurs when warm air meets a cold surface: the moisture in the air cools and condenses on the surface in tiny droplets of water. Most commonly, condensation occurs on the inside of single glazed windows, but it can also occur on walls and within the fabric of buildings.

Condensation is a problem that has a number of causes:

- It can be an indication of the internal temperature being too low for the internal humidity level. This can happen in an unheated room that is not being used. The insulation of the whole building helps to solve this problem by evening out the temperatures internally, making it easier to keep the extremities of the house warm.
- It can also be an indication that humidity levels are too high and that water vapour produced elsewhere in the home is not being vented. The simplest way of avoiding this problem is to use mechanical extract fans connected to a humidistat, in both the bathroom and the kitchen. The humidistat will switch on when humidity levels exceed a predetermined limit. It is also a good idea to keep closed the door of a room where there is a source of water vapour (see ventilation section under draught proofing).
- Condensation in an insulated house can also be an indication of cold bridging: where a non-insulated part of the construction protrudes through the well-insulated part and causes a surface in the interior to be at a much lower temperature than its surroundings. If the humidity level is high enough, condensation will form on this cold surface, possibly causing problems of deterioration. Cold bridging occurs most commonly at windows and where brick partition walls meet internally insulated external walls.

Cold bridging at windows

With double-glazed windows, the problem of cold bridging usually only arises with metal windows. Metal, being a very good conductor of heat (and thus a poor insulator), conducts heat much more quickly to the outside through the frames. If metal (either aluminium or steel) is to be chosen as a material for the frames then there needs to be what is called a thermal break incorporated into the frame. This is simply an insulating material sandwiched into the construction. Alternatively, wooden windows can be used—which is the more ecological solution anyway.

Cold bridging at partition wall junctions

This occurs with internally insulated external walls—any partition making a junction with the insulated wall bridges the insulation and loses heat much more rapidly at this point, producing a cold internal face adjacent to the insulated wall. This problem can be partly overcome by returning the insulation on the section of partition adjacent to the corner. If it is a plaster wall then the plaster can be removed at this point to be replaced by insulation.

To sum up: although ventilation and heating play a role in preventing condensation, good insulation correctly installed provides the most energy-efficient method of combating condensation.

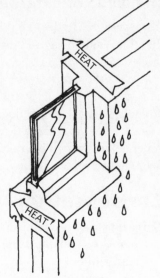

Cold bridge at window reveal.

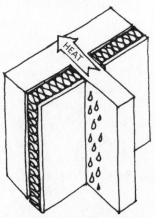

Cold bridge at partition.

Insulating your roof

Your loft is either lined or open such that you can see the whole structure. The latter is still the most common, although more and more people are undertaking loft conversions.

Unlined lofts

This is the insulation of the ceilings of the uppermost floor from above. If your loft space is not used except for water storage tanks and services, then you have a simple job to insulate between the joists. Most people in this situation already have some form of loft insulation, but if you are installing from scratch lay 150mm or 200mm of insulation depending on what you have space for. If you have less than 100mm, it is worth topping it up; otherwise work on other areas of the house first and come back later when you can afford to. The main principle to follow here is to cover over any pipes or tanks that contain water, as the air temperature above the insulation can go below freezing.

Ventilation to be maintained at eaves

Ensure water tanks are lagged, with ceiling below left uninsulated

Loft insulation at ceiling level.

Lived-in attic space or lined lofts

If your attic space is habitable—whether it is heated or not—it is advisable to improve its insulation. In this case, we are concerned with insulating the roof rafters and not the joists. There are different ways of doing this, depending on the construction. Insulation can be fitted:

- Below the rafters
- Between the rafters
- Above the rafters or
- Some combination of the above

1. Below the rafters

If the rafters are open and unplastered, it is a simple matter to place sandwich insulation boards on the inside and fix plasterboard underneath. This method of insulation reduces the room height and may inhibit the thickness of insulation possible.

You may want to add insulation between the rafters as well, to increase the thickness of insulation. In this case, it is necessary to leave a gap of 50mm between the insulation and the roofing felt, to allow the rafters to breathe. This may well determine the thickness of the insulation possible. There is an advantage if at least one of the insulation boards is covered

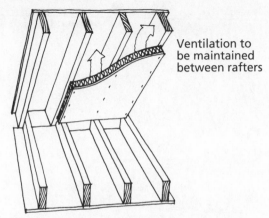

Ventilation to be maintained between rafters

Loft insulation below rafters.

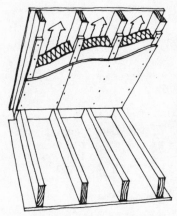

Ventilation to be maintained above insulation

Depth of rafters increased to allow greater thickness of insulation

Loft insulation between rafters.

with a layer of aluminium foil to add to the effectiveness of the insulation. It is also necessary to place a vapour-proof barrier between the insulation board and the plasterboard. If your attic space is already lined, and you do not want to go to the trouble of removing it (and headroom is not critical) the insulation could be placed directly on the underside of the lining.

2. Between the rafters

This is the most usual place to install the insulation: the main problem here is often that the rafters are not deep enough to accommodate a reasonable thickness of insulation. However, with this method it is possible to increase the depth of the rafters by adding a lath underneath each of them to whatever thickness is necessary to allow the 50mm gap between the insulation board and the roofing felt. For example, if you have a rafter of 75mm depth and you want to add 50mm of insulation board, you could attach a lath of about 25mm underneath. The same would apply with the silver foil and the vapour-proof barrier.

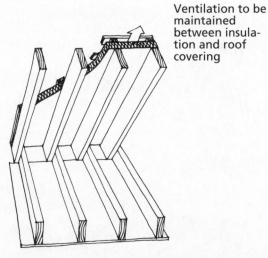

Ventilation to be maintained between insulation and roof covering

Loft insulation above rafters.

3. Above the rafters

This particular option is only worthwhile if you are having your roof renewed. It is also the most complicated method because it almost certainly involves the redesign of the eaves to allow for the added thickness of insulation. The basic principle is to strip the roof of tiles and tile laths, to lay a rigid insulation board on top of the rafters (with a vapour-proof barrier in between) and then lay a building board or roofing laths on top of this to attach the tiles or slates. It may be necessary to employ an architect or someone who can competently produce details of how the construction at the eaves can be altered. This type of insulation enhancement is likely to become much more common in the future, and with routine construction details being developed for the eaves, it will allow almost any thickness of insulation to be installed on the roof.

Wall insulation

Most external walls in Britain are of brick or masonry construction. The insulation of other types of construction can broadly follow either the same pattern as that for the roof construction which has already been described, or the construction described below.

There are three possibilities for the insulation of external walls:

- Insulating the internal cavity
- Insulating externally—the most effective form of insulation
- Insulating internally—the most involved

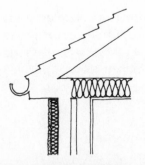

External insulation with rendered finish.

Cavity wall insulation

More heat is lost through the walls than through any other part of the house. Yet only 20% of homes with cavity walls have cavity wall insulation. This is a simple measure to undertake, with very little internal or external disruption. It is advisable to have it installed by a professional contractor—it usually takes less than a day for the insulating material to be injected into the walls from the outside. This is therefore a very obvious measure to take if you can, as there is no internal or external upheaval. Various insulating materials can be injected into the walls, including mineral wool.

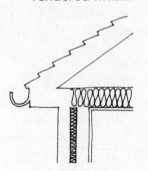

Cavity filled with blown insulation.

External insulation

As was pointed out earlier, the most effective place to install insulation is on the exterior of your property. It keeps the whole fabric warm and dry, and the contained walls provide the

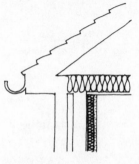

Internal dry-lining.

thermal capacity which can even out extremes of hot and cold. However, this is also a professional job and can be expensive. The biggest design problem occurs at the windows, and here again it is important to have some design advice and not necessarily leave it to the contractors who are installing the insulation. This solution is particularly appropriate if your house is rendered, but may also be so if the general quality of the outside is not of a high standard.

Internal insulation

There are many situations where it may not be appropriate to place the insulation on the outside of a building, such as in a conservation area or where the brickwork or stonework is particularly attractive. In these cases (assuming there are also no cavities between the walls) the insulation needs to be installed on the inside of the external walls.

This can be done in two ways:
- By attaching rigid board insulation to the existing interior wall, then covering it with plasterboard (some plasterboard has insulation already attached).
- By building a new stud wall against the inside face of the masonry and placing fibreglass bats (mineral fibre glued together in a semi-rigid quilt) between the studs and covering the studs and insulation with plasterboard. Architraves around doors and windows need to be removed and replaced after re-lining.

If you decide to undertake the internal insulating of your external walls then you have a choice as to whether you do it all at once (the cheapest way, if you are using a contractor) or installing the insulation room-by-room as you redecorate.

Window insulation

Once you have insulated the roof space and walls it becomes essential to double- or even triple-glaze your windows, as they will now be the main cause of heat loss. If you are insulating your walls you may well want to consider the windows at the same time.

Principles of double-glazing

The main principle behind standard double-glazing is that of trapping an insulating layer of air or other insulating gas between two panes of glass. The distance between the two panes is important because if the the two panes are too close there is increased conduction of heat across the gap, and if they are too far apart then there are convection currents set up between the two panes of glass, which will decrease the insulating effectiveness. The ideal gap is about 12mm. Increasing this gap does not significantly improve its insulating properties, but does improve the sound-proofing. There are two other ways of increasing the effectiveness of double-glazing:

Improving the greenhouse properties of the glass

The normal properties of glass are to be transparent to sunlight (the visible spectrum) and reflective to the longer infra-red wavelengths of heat from inside. This reflective quality, which keeps the heat in, can be further enhanced by the use of special coatings on inner panes.

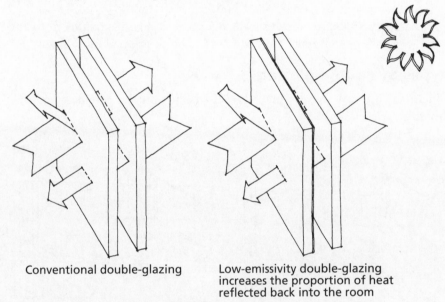

Conventional double-glazing

Low-emissivity double-glazing increases the proportion of heat reflected back into the room

Improving the insulating properties of the frame

The insulating properties of the frame itself are important and can help increase the overall insulating effectiveness and prevent condensation arising from thermal bridging. Solid metal frames conduct heat quickly and should be avoided unless they contain a thermal break. Plastic has the advantage of low maintenance in the short run, but its long term durability is unknown. Timber provides the best all-round ecological solution if the timber is well seasoned in the first place and is kept protected.

Improving the insulating properties of the gap between the panes

The insulating layer between the two panes of glass can be improved by either producing a vacuum between the two layers of glass or substituting the air for a light gas such as argon or krypton. Producing a vacuum is an ideal way of cutting out the conduction between the two panes of glass. In practice this is difficult to attain and is only possible in factory sealed glass envelopes that are expensive and limited in size. However, factory-sealed units containing argon or krypton are becoming much more common.

The benefits of double-glazing

There are several benefits of double-glazing besides saving energy:
- The amount of condensation on the panes will be reduced or eliminated.
- Noise from outside will also be reduced, especially if effective draught-proofing seals out air-borne sound.
- There will also be a significant decrease in down-draughts from the windows, which will allow radiators to be positioned more freely.

Types of double-glazing

There are three ways in which double-glazing can be added to existing housing.

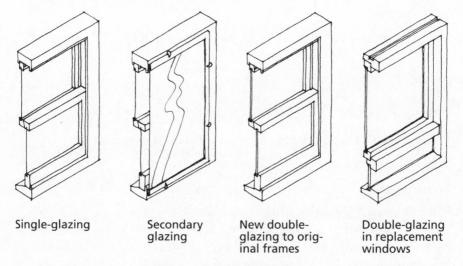

Single-glazing   Secondary glazing   New double-glazing to original frames   Double-glazing in replacement windows

Types of double-glazing.

- Secondary glazing: this is a second pane of glass or plastic, either fixed or openable, that is installed on the inside of the existing window. Increasingly these systems have come to look like a second window, including a second frame.
- Sealed units: these consist of two panes of glass which are sealed in a workshop or factory with the use of a separating edging strip. These units can then be used to replace the panes of glass in the sashes or fixed

windows. Thicknesses of these vary from 10 to 20 mm, giving a choice of thinner units to fit narrower sash or mullion widths.

- Replacement windows: these are whole windows that replace the existing window, frame and all, incorporating sealed units.

How do you decide what type of window to choose? The three most important considerations are:

- The state of the existing windows and whether double-glazed units can be incorporated into the existing sashes.
- The aesthetic or architectural quality of the existing windows.
- Cost.

The state of the existing windows and aesthetic quality

Your windows may be in a poor state of repair, being partly rotten (if wooden), or corroded (if metal). The decision whether to replace completely or repair can be a difficult one, but ultimately the expenditure and savings in terms of energy, money, time and materials need to be balanced up. Although almost any timber window can be repaired (and this should be the preferred option),

Original window in timber with small-paned sliding sashes.

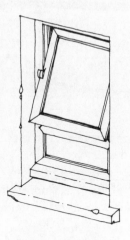

New window in unsympathetic design and materials.

New window in design and materials sympathetic to original.

it is sometimes the most sensible step to start again if you can considerably increase the insulating value of the window by doing so. What will often be the deciding factor will be the aesthetic quality of the existing windows compared with that of the replacements.

Other ways of improving the insulating properties of your windows

There are many ways that windows can be insulated at night, or when a particular room is not being used in winter:

- Curtains: there are many ways that curtains can be made more effective. You can use thicker material, or even a quilted material with an insulating filling. You can add a reflective covering to reflect heat back into the room, and you can make sure that escaping down-draughts from between the window and the curtains are reduced by:

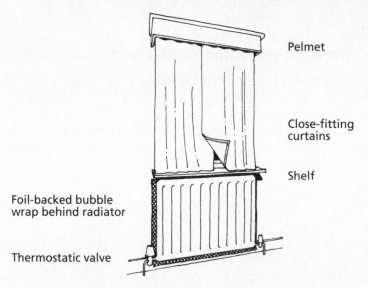

Pelmet

Close-fitting curtains

Shelf

Foil-backed bubble wrap behind radiator

Thermostatic valve

- using a pelmet which fits well around the top of the curtain
- ensuring that the sides of the curtain connect as well as possible to the sides of the frame and to themselves in the middle
- providing a shelf or sill for the seam of the curtains to lie on
- ensuring that if a radiator is fitted below the window, the warm convection current from the radiator does not go up behind the curtain (either tuck the curtain behind the radiator or fit a shelf).
- Blinds can be made to fit in slides at the side of windows, so they are easier to seal than curtains. The bottom of the blind can simply rest on the sill and a simple flap can contain the top. Blinds, being more rigid than curtains, can more easily be treated with special coatings or incorporate a thin insulating film.
- Shutters—these can be designed to be both insulated and tight-fitting, and hinged so that they fold back and to the side when not in use. However these will almost certainly need to be specially made.
- Pop-in insulation panels—these are a cheap and effective solution. A

ridged sheet of insulation, prefer-
ably one that is a sandwich of
aluminium foil and card (in order
to protect the foil), is cut to size
and fitted to exactly the inside of
the frame and the side facing the
room covered with fabric. The
main disadvantage is the extra work
required to place them each time
you wish to insulate the window.
One solution is to have them
hinged at the top so they can swing
up out of the way and be attached
to the ceiling by a hook.

- External security and insulation
blinds. These are made first and
foremost for security. However
they have the advantage that they
are placed on the outside of the
house and so do not interfere with
internal arrangements of rooms.
They would be worth thinking
about if you are having external
insulation fitted, or you have a real
security problem in your neighbourhood, or the facade of your
house has little aesthetic significance.

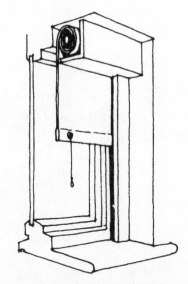

Insulated blind running in channels.

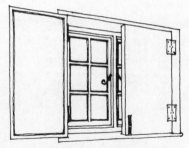

Close-fitting shutters.

Ground floor insulation

If you have suspended wooden floors, it is relatively easy to insulate them by
laying either mineral fibre, rigid or loose-fill insulation between the floor
joists (see p.68). This is sometimes possible to carry out from below (if there
is adequate height in the crawl-space), or from above by removing the floorboards. If the floor is solid it will be necessary to either relay the whole floor
or raise the floor level by adding a layer of insulation on top of the existing
floor and covering with tongue-and-groove boards or chipboard (see next
page).

Basement insulation

Basement walls can be insulated on the inside with either rigid boards or

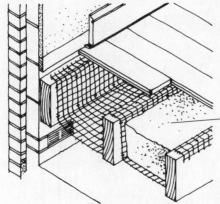

Skirting board on compressible strip

Mineral fibre insulation suspended from floor joists by plastic netting

Insulation of suspended ground floor (1).

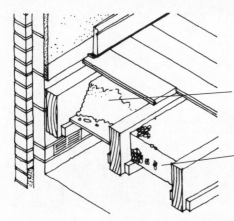

Skirting board on compressible strip

Loose fill insulation supported by boards on battens nailed to floor joists

Rigid polystyrene insulation supported directly on battens

Insulation of suspended ground floor (2).

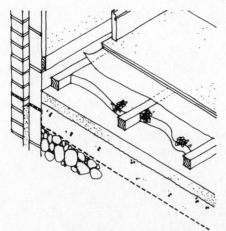

Chipboard flooring

Vapour barrier

Polystyrene insulation between timber battens

Existing solid floor

Insulation of solid ground floor (raising its level).

fibreglass batts, using the methods described above. Another option is to attach moisture-resistant insulation to the outside of the wall below ground level. This can be done by digging out and fixing the insulation to the external wall one metre below ground level. At the same time it is possible to water-proof the outside of this wall and carry out any damp-proof course work that may be necessary. It is worth remembering that below the surface, the earth is a good insulator so the levels of insulation required are considerably less than for above ground.

Priorities for action

Upgrading the insulation of your home is undoubtedly the most involved and costly of all the measures in this book. For most people it is not feasible to carry it out all at once. The important thing is to undertake the work step by step, obtaining the best advice you can beforehand.

1. If your roof space has not been converted or insulated, undertake this first with 150-200mm of cellulose, glass fibre or mineral wool. If your existing insulation is less than 100mm then top it up with an extra 100mm. If your attic is converted see pp. 59-61.

2. Insulating the walls of your house is the next most important measure to undertake. If you have an uninsulated cavity, have a reputable contractor carry this out. If your walls are solid, look at the wall insu-lation subsection on p.61 to help you decide which is your next option to choose from. If you are applying external insulation you will almost certainly require a specialised contractor. If you are applying insula-tion internally, try and undertake this work when you are having the house decorated.

3. Double-glaze your windows at the same time as you insulate your walls if possible, and use low emissivity glass to make it as effective as triple-glazing (refer to the section on double-glazing).

4. Ground floor: if you have a suspended wooden floor, insulate between the joists either from underneath or from above by taking up and then replacing the floorboards. Take this opportunity to properly draught-proof the floor and ensure that your underfloor space has through ventilation.

5. Basement or cellar (if you have one): concentrate on insulating any parts of the external walls that are above ground, either internally or externally. Damp-proof and improve ventilation before any extensive insulation work.

Space heating

Heating our homes, including the heating of hot water, accounts for the vast majority of energy that is used domestically, and represents one of the greatest wastes of energy that we at present indulge in. Many of us have outdated, oversized and inefficient systems that are not making the best use of the energy delivered to them. After you have draught-proofed and insulated your home, you will almost certainly find that your heating appliance or appliances are oversized and if you have done the job well, you will be able to save up to 75% of the energy expended previously on heating. This chapter looks at the factors that affect our heating systems and points towards ways of making the very most of the systems we have. These include:

- Identifying the different sources of heat within our home and how we can conserve them
- Choosing an appropriate way to update our system
- Maximising the potential of the total system
- Choosing and using controls effectively
- Maintaining the system for maximum efficiency

Conserving the heat within our homes

Our space-heating appliances are not the only sources of heat within our homes. Other sources of heat are:

- Lights
- Cookers
- Hot water
- Refrigerators
- Colour televisions
- Washing machines
- Dishwashers
- Clothes dryers

The heat from all these sources should be borne in mind when considering our overall heating requirements. If our homes are well insulated, all this heat can be conserved and these additional sources then begin to provide a larger contribution to heating the home. Kitchens for instance often hardly require any additional heating. However, the list above does not include two very important sources of heat:

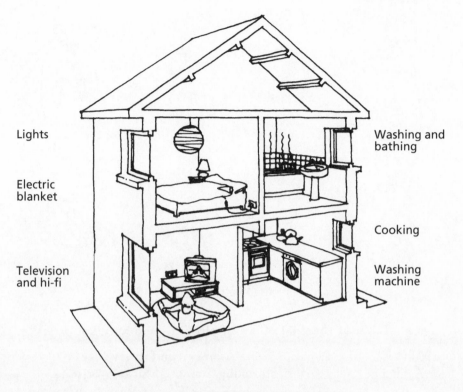

Lights

Electric
blanket

Television
and hi-fi

Washing and
bathing

Cooking

Washing
machine

Other sources of heat in the house.

- Light and heat from the sun
- Heat from our bodies

Making the most of the free solar energy that falls on our homes is the subject of the last section in this chapter. But what is the relationship between the heat we produce in our bodies and the thermal environment within our homes? The main purpose of heating our homes is to keep our bodies warm. A secondary purpose is to keep our homes dry to preserve materials, but this could be done more efficiently by other means. What is not generally appreciated is that heating our environment does not heat our bodies; it only slows the rate of heat loss. The heat generated by the human body has to be dissipated to keep its temperature from rising unacceptably. In other words, to remain comfortable we require surroundings that are cooler than our bodies. To a large extent we are the source of our own warmth: we have ways of increasing the heat we generate within ourselves through exertion, as well as ways of insulating ourselves to remain at a comfortable temperature by wearing clothes.

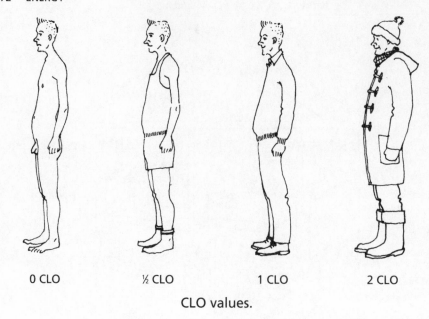

0 CLO   ½ CLO   1 CLO   2 CLO

CLO values.

Table of thermal effect of clothing and activity

Clo Value		Average comfort level °C			
		Strolling	Standing	Sitting	Resting
0	Nude	21	27	28	30
0.5	Light Clothing	15	23	25	27
1.0	Normal Clothing	8	19	21	24
1.5	Heavy Clothing	0	14	18	21
2.0	Very Heavy Clothing		10	14	18

Ref: Peter Burberry—Building for Energy Conservation (1978)

From the above we can see how we keep ourselves warm and the way we use clothes to moderate the effects of both surrounding temperature and exertion. We do the same at night when we put more insulation on our beds when we are cold. If we are looking for ways of housing ourselves ecologically, it is these more natural ways of keeping warm that we must return to. The temperatures we set for ourselves in winter are gradually creeping up as we get used to higher temperatures, shedding more clothes and living a more sedentary way of life. Many of these trends are not healthy for us or the planet.

The importance of these factors becomes apparent when we realise that for every degree rise (°C) in temperature that we turn the thermostat up, we increase the fuel burned by approximately 10 %, resulting in increased CO_2 emissions. We should not forget, however, that as the body gets older and its metabolism less efficient, it requires a higher ambient temperature.

So the message is one we knew already: we can save considerable amounts of heat by wearing more clothes and sitting around less. When we are in bed we certainly don't require any heating, as it is much more effective to buy a good duvet. If we do require to sit and work for longer periods, we need to use the room that stays warmest or is the easiest to heat.

Different ways of heating

You will already have a heating system in your home (80% of homes in Britain have central heating or storage radiators). Most of these systems have been designed to cope with the higher heating load of a badly draught-proofed and insulated house. Many of them have the capacity to heat the house on the coldest day of the year to 21°C and it is arguable that this is necessary. Many heating systems work most efficiently when they are working at maximum load, which means that if the system is designed for the coldest day it will be working less efficiently on all the other days. This problem is being solved with the latest designs of boilers and their controls.

Since we are looking at the ecological upgrading of an existing system we need to review the situation with the assumption that you have draught-proofed and insulated your home to as high a standard as you can afford. The most ecological sources of energy have already been identified. The best system for your particular home will depend on the characteristics of the building and your lifestyle. The alternative systems to choose from are as follows:

- Full central heating
- Partial central heating
- Individual space heaters where required
- Reliance entirely on incidental sources of heat

There are various reasons why you may not require a full central heating system. Here are some of them:

- If your house is small, compact and well insulated
- If your house receives sufficient passive solar energy
- If your house has an Aga or equivalent range
- If your house has a large ceramic masonry heater
- If you intend to concentrate your activities in only one or two rooms

We shall start by looking at full central heating, and work back from there.

Full central heating

The vast majority of central heating systems use water as the medium to transport the heat from the central boiler to where it is required. There are many other types of central heating such as hot air systems, underfloor heating and steam systems. However I shall deal mainly with 'wet systems' as it is the system you are most likely to have.

A wet central heating system consists in most cases of the following parts:

- The boiler that burns the fuel and transfers the heat to water
- The pipework that transports the hot water usually with the help of a pump
- The radiators that transfer the heat from the water to its surroundings
- The controls to ensure that the right amount of heat is delivered to where required at the right time

If central heating is already installed, the first question to ask is whether or not it is the right size system. If you have improved the insulation as indicated in the last chapter then, as has already been mentioned, it is likely that the system will be oversized and the boiler size can be reduced when you next replace it. You may also want to reduce the size or number of the radiators if this will enable the more efficient use of any of your rooms.

If you don't have central heating, you may well be considering whether it is right for you. You need to make a realistic assessment of the costs of your existing system. For example, you may have individual heaters using existing chimney flues which could be resulting in considerable heat loss. If you have a relatively large house and you need to keep the majority of it warm most of the time, it is probably sensible to go for a high-efficiency gas central heating system and for the smallest boiler size compatible with the insulation.

A partial central heating system

If however you have a medium-sized two-storey well insulated house, it is certainly not essential to install full central heating in order to ensure an adequate distribution of heat. Evidence suggests that if the ground floor is kept warm through central heating, sufficient heat finds its way upstairs by natural air movement upwards, encouraged by the stack effect (see p.49) and conduction through the ground floor ceilings to provide an acceptable temperature in upstairs bedrooms.

Even if you have central heating, it is probably a good idea to place a heater with some radiant output in the main living room to allow a quick warm-up of the principal room. It is useful in autumn and spring to be able to warm one room without having the whole system running.

Individual space heaters and alternatives to central heating

There are many different types of unit space heaters. Some of the most efficient now are the gas wall heaters with balanced flues. There is also the possi-

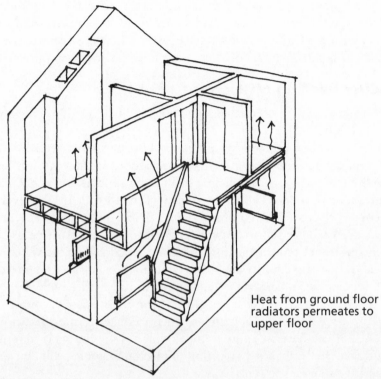

Heat from ground floor
radiators permeates to
upper floor

Partial central heating.

bility of an efficiently designed woodburning stove, if you are willing to spend the time tending it to ensure that it burns correctly. Many people have been persuaded into heating their homes with electric storage heaters because of cheap rate tariffs—which bear very little relationship to the amount of CO_2 produced—and the cheaper initial cost of installation. However, as indicated in the fuel section, heating your house wholly by electricity is not ecologically sound and should certainly be avoided. Electric heaters can be justified only if they are used *sparingly* as a top-up heater to be taken on occasion to anywhere in the house and used in a very localised way for a *limited* period of time. The most effective heaters for this purpose are the small fan heaters with built-in thermostatic control. However, think first of wearing more clothes!

Increasing efficiency

Choosing a new boiler

If you either have an existing system to be upgraded or have decided to opt for a whole new central heating system, you will have to give careful consideration as to the type of boiler. This is after all the heart of your energy-efficient home, and making the right choice can make a big difference in terms of CO_2 production and money saved. We look eagerly at the fuel efficiency (m.p.g.) of cars but rarely do we look at the comparative costs of different boilers, which could be using far more fuel than our car.

Boiler heating efficiency

How much of the potential energy in the fuel that is fed into the boiler ends up in the heated water? This proportion, if expressed as a percentage, gives a measure of how efficient your boiler is. Usually it is between 60% and 90%. The most efficient boilers in terms of low carbon dioxide emissions are natural gas boilers, and the most efficient of all is the condensing type. Unlike most boilers, this maintains a high efficiency even at low loads. Its 'cycling' efficiency is close to 90% compared with 70-75% for a new conventional gas boiler. To achieve this increase in performance, the waste flue gases, which contain steam from the burning methane, are cooled by the water returning from the radiator circuit. The steam condenses, giving up its high latent heat. The cooler the return water temperature, the more efficient the boiler will be.

There are many other types of boiler on the market but none approach the gas-fired ones when it comes to energy efficiency and ecological performance. If you have an existing boiler which is still relatively new and is not vastly oversized, then pay attention to maintaining it at peak efficiency until you have a chance to replace it.

The circulating system

Most circulating systems have a flow pipe carrying hot water from the boiler and a return pipe taking it back. Each radiator is connected to both. This ensures that all radiators get hot water direct from the main flow pipe. In a single pipe design, used in some older systems, the water flows from one radiator to the next and only the last radiator in the circuit is directly connected to the return pipe. Older systems also used large diameter pipes because the water flowed in a convection loop (or gravity system) without the aid of a pump. Since the introduction of electrical pumps the trend has been to smaller and smaller bore pipes under greater pressure. Small bore pipes are generally either 15 or 22mm in diameter (most systems use both sizes for different flow rates according to how much water they need to carry). Microbore systems have yet smaller pipes again—between 6 and 12mm in diameter—and each radiator has its own flow and return pipe connected to a central manifold. These systems can be very neat and avoid the need for much damage to the existing structure during installation. Where pipes are passing through basement or underfloor areas, lofts or unheated spaces, they should be lagged. This is particularly important for flow pipes which take the hot water to the radiators. The exceptions are where you would like to make use of the heat from a particular section of pipe—through a cupboard, for instance, to keep it aired.

The radiators

Radiators are where the heat is actually delivered and as such are more important elements than we might think. These are some ways to make the most
of them:

- Each radiator needs to be 'balanced' with the rest of the system in order
  to ensure that water is flowing through all the radiators equally by
  adjusting the valves at either end.
- Radiators collect air that has become dissolved in the water and need
  to be bled at least at the beginning of each heating season and more
  often if the water has had to be changed.
- Fix a shelf above the radiator—this will help to throw heat back into
  the room that would otherwise go straight up the wall to the ceiling.
  If fitted beneath windows so that the bottom of the curtain can rest on
  the shelf, this will prevent warm air from being lost between the window
  and the curtain (see illustration on p.66).
- Place an insulated reflector between the radiator and the external wall—
  this reflects the heat back into the room.

Improving the controls

Probably the most important way in which you can save wasted heat is to
have adequate controls and know how to use them. Controls will keep the
rooms only as warm as you actually need them, and be able to turn the system
on and off as required. They can, for example, turn the system off during
the night when it is not required, and off again during the day when there
is a higher external temperature. These controls need to be sophisticated
enough to be able to perform what is required and yet simple enough for us
to understand them. They can be tailored to the needs of the house and its
occupants.

What are the different types of control? There are the thermostats which
are situated wherever it is necessary to monitor and control the temperature
of water in the system or the air in a room or even outside the house. Then
there are the timers and programmers which send signals for systems to be
shut down or opened up at pre-set times.

Thermostats

The Room Thermostat

This monitors the temperature in a chosen room and switches the boiler on
and off according to the temperature in the room. With such a system it is
very important to place the thermostat in the right place so that it is indica-

tive of the temperature of the whole house. In most cases it is fixed to the wall of the most-used room, away from direct sunlight and draughts or any other source of heat or cold. The main problem with this system is that, by itself, it does not control the temperatures in rooms other than the one in which the thermostat is placed. This can be overcome to some extent by fitting thermostatic radiator valves to each radiator in all the other rooms.

Thermostatic radiator valves (TRVs)

These are a combined thermostat and valve which work very effectively to turn the water running through the radiator on and off according to the temperature in the room. They look a bit like taps and can be set at different numbers that indicate higher or lower temperature settings. It is a good idea to use these valves along with a room thermometer in order to know what temperature will be achieved with each number setting. One of the great advantages of these valves is that they can be turned to any setting required and also turned on and off like lights as heat is required.

Boiler thermostats

These are an integral feature of the boiler: a dial is usually positioned on the outside along with any other controls. Most have numbered settings rather than temperatures. Generally it is better to keep the boiler setting high as most run more efficiently at high temperatures; but consult a heating engineer to find out the most efficient temperature at which to run your boiler.

Timers and programmers

These are simply a means of switching the boiler on and off at preset times which can be one of the most important ways of saving energy. Programmers are becoming more sophisticated so that you can set them to a different programme every day of the week. Together with good insulation, which slows down the cooling of the hot water tank and speeds up the heating process, it generally means that the length of time the boiler needs to be working is considerably less.

Advanced controls

There are more and more new developments in microchip controls for heating systems that will become increasingly available to the domestic consumer. These can be used to co-ordinate many diverse functions to gain the greatest efficiency out of the system. An example of the sort of function these perform is given by an optimiser which allows you to set the actual times you require heating without having to make allowances for how long the house takes to heat up. The optimiser will automatically monitor your system over a week or so and learn from experience the time that needs to be allowed for pre-heating. Then it will turn the boiler on in time to have the house warm when you need it.

If you want to re-evaluate the overall efficiency of your system it is very important that you identify someone who has the latest information and knowledge. Many people rely on plumbers for this information, but they are often too busy to keep up with the latest developments. Try hard to find someone who can analyse your overall needs and who is interested in energy-efficient systems. You will also require a specialist to maintain your system in peak working order.

Priorities for action

1. If buying a new heating appliance then buy the one that will generate, directly or indirectly, the least amount of CO_2. In most cases that will mean using either natural gas or bottled gas. Remember that good insulation is a key part of an efficient system!
2. When deciding the best heating system for your home, consider a partial central heating system if suitable for your pattern of living (see p.74).
3. Undertake measures to maximise the efficiency of your system—this means putting some thought into how its efficiency is affected by each separate part (see pp.75-79).
4. Use energy-saving measures in conjunction with your radiators (see checklist on p.77).
5. Obtain the very best advice you can find to help you plan the most appropriate system of controls for your particular house and lifestyle. These controls should help ensure that your boiler generates only the right amount of heat, where you want it and when you want it.
6. Buy gas equipment with automatic ignition. Pilot lights waste substantial amounts of energy.
7. Maintain your boiler at its most efficient. Read the maintenance manual and carry out regular services.
8. Understand the controls you have and use them. It may be worth obtaining advice on how to obtain the most from your system.

Water heating

There are several different ways to heat water for domestic use. You will probably already have a system installed and be more or less pleased with its performance. It is the function of this chapter to help in reviewing the choices available and deciding whether you need to consider renewing your system.

An average family of four uses 700—1200 litres of hot water a week. The energy needed to heat this water often amounts to about one fifth of the total energy used (and it can be a higher proportion in a well insulated house). Besides choosing the right source of energy, there are ways of using less water more effectively. Some of these are also discussed in two other chapters in the book: WATER on p.130 and APPLIANCES on p.94.

The first point to understand about water heating is the difference between instant systems and storage systems. These are the two basic approaches, and they are suitable for different circumstances. Instant systems are those that heat water as required, such as gas multi-points or electric showers; and storage systems keep hot water available all the time, such as cylinders heated by electric immersion heaters or by the central heating boiler.

Choice of energy source

In addition to these two main systems there is the matter of energy source. Here there is a choice between the following, given in order of ecological soundness:

- Solar energy
- Gas
- Oil
- Coal
- Electricity

In fact if you are heating all your water electrically, replacement of your system should be a high ecological priority. The problem is that very many people do heat their water electrically, because of its convenience; even if we don't heat our main supply of hot water electrically we use washing machines, dishwashers, showers and kettles that do. Once we become aware of the environmental impact of all this usage we have a choice about how often we use these appliances and also begin to find alternative ways of carrying out their functions.

At the top of the list is solar energy, which has many characteristics opposite to those of electricity. It is intermittent and dependent on the seasons and the weather. It is expensive to install and cheap to run. It requires much care in its design, installation and operation. However it is essentially unpolluting and can complement a storage system that is run by a central heating system boiler in the winter.

An assessment of the relative merits of gas, oil and coal has already been made at the beginning of this chapter (see **SOURCES OF ENERGY** on p.40).

Choosing between an instant and a storage system

The choice between an instantaneous system and a storage system will depend on a number of factors: the size of the house, the number of people using the system, how frequently the system is used and how uneven the usage is. The choice of an instantaneous system is indicated if:

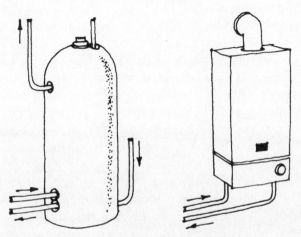

Storage heater (left) and instantaneous heater (right).

- The house is small.
- The number of people using the system is small.
- The system is used unevenly, sometimes requiring large amounts of water but most frequently small amounts.
- There is little requirement for multiple use i.e. when two or three people want to use hot water at the same time.
- There is no space for a storage cylinder.

The choice of a storage system is indicated if:
- The opposites of the above are true.
- There exists an efficient boiler that can also be used to heat a storage cylinder.
- There are plans to install a solar panel or a heat recovery system with a heat pump.

Going through the above list and finding out how many indicators you have for each system should give you a good idea of which one is most suited to your needs.

Instantaneous systems

Instantaneous systems are essentially appliances which heat the water as it passes through, producing hot water on demand. The heating is switched on or ignited when water begins to flow through the unit. There is usually a minimum flow that needs to take place before the heating is activated, to avoid overheating.

Two types of instant system are available: gas and electric. Gas units consist of multi-point heaters and combination boilers that heat water for central heating as well. Electric systems include those that heat water for showers, and miniature units that heat small amounts of water for hand-washing at remote locations. Much has been said about the heating of water with electricity and this would include the shower units that have become so popular. It is quite possible to run showers from either multi-point or combination boilers if the plumbing is arranged correctly. The use of mini-electric heaters for warming water to wash hands are useful only at the end of long pipe runs where more energy would be wasted running the tap than heating the water electrically.

The main advantages of instant systems are that (a) there is no need for a storage cylinder, which is constantly losing heat; and (b) only the water that is required is heated, which prevents waste when there is intermittent demand for hot water. Storage systems will cool down after a bath so that it then takes time for the water to heat up again. Their main disadvantage is that only a limited amount of water can be drawn off at any one time.

Some gas combination boilers take longer than others to fill a bath, especially if hot water is being drawn off from elsewhere in the house. However, gas-heated instantaneous hot water is generally the most appropriate for a small house and could also be for a larger one if the conditions are right (see above).

Storage systems

The storage system depends on keeping water hot in a tank or cylinder. The water is heated in the cylinder itself; this can be done in several different ways. An electric immersion heater, acting like an oversized electric kettle, is one way.

Alternatively, a calorifier or heat exchanger, often in the form of a coiled tube, is used to transfer heat from water outside the cylinder: most commonly a boiler. Other ways this water can be heated are through the backboiler of a solid fuel fire, a boiler in a wood stove, an Aga or of course almost any central heating boiler. If you use your central heating boiler you will require a programmer that can control the two systems together.

Needless to say, the immersion heater should be a water heater of last resort and if at present you rely entirely on this form of heating then it is worth thinking of replacing it with a multi-point or a combination boiler.

The most important point of all if you are using a storage cylinder is the insulation. Cylinders are losing heat continuously, day and night, and this loss can account for a large amount of energy usage during the year. Most cylinders come ready insulated with foam. If you are ordering one ask for an extra layer of foam. There is also nothing to stop you packing more insulation around the cylinder when it is installed.

Traditionally, storage cylinders required a cold water storage cistern. However in recent years water by-laws have allowed storage cylinders to be fed direct from the mains, doing away with this requirement. This is called an unvented system and is widely used in continental Europe.

Storage cylinders and cisterns are very heavy indeed when filled with water. It is thus important that these are well supported in two ways:
- The structure on which the tank stands, and
- The cylinder or tank itself needs to be supported properly across its base to prevent rupture.

The main advantage of storage cylinders is that they allow multiple usage at the same time. There is also the possibility of using heat-recovery technology to recover heat from waste hot water and waste hot air. The Eco House in Leicester uses a heat pump to recover heat from waste hot water, and in some Swedish super-insulated houses heat is recovered from ventilation exhaust air to meet all the water-heating needs.

Hot-water cylinder thermostat

This is a small thermostat that may be fitted to the outside of the hot-water cylinder (not to be confused with an immersion heater). Setting it at, say, 50 °C will ensure that the water in your cylinder is only heated when the water temperature is below that setting. It avoids the problem of your tap water becoming scalding hot. It also reduces the unnecessary operation of your boiler.

Solar water heating

A little is included here on solar water heating as it directly relates to this section. Solar heating systems in the UK cannot provide all domestic hot water needs around the year, because of our latitude. However they can be used to pre-heat water before it enters a conventional cylinder heated by a gas boiler, for instance. This system complements the boiler, which is doing most of its work in the winter, as a properly installed solar water heater should provide all summer hot water requirements.

There are two systems: direct and indirect. In direct systems the heat transfer liquid passes straight to a hot water tank. In indirect systems there is no direct contact between the transfer fluid and the hot water system. The solar-heated water/fluid is circulated from the storage tank through a heat exchanger coil immersed in the hot water cylinder. More information about solar systems is given in the SOLAR ENERGY chapter on p.102.

Hot water conservation

As with space heating, conservation is cheaper than the cost of the energy itself, or of a new installation. There are two basic means of conservation. Firstly you can heat the water with an energy-efficient appliance; secondly, you can conserve energy by:

- Insulating the pipework and cylinder.
- Reducing the temperature at which hot water is stored by using a cylinder thermostat and setting at the lowest temperature that will give you the hot water you need.
- Saving on the actual amount of water used by having spray fittings on taps for washing hands. It is also worth using an adjustable shower head since it is then possible to vary the amount of water being delivered according to how much hot water is available.
- Insulating the bathroom or shower to reduce heat loss while bathing and showering.
- Recycling the waste hot water through a heat pump to extract the heat before it is lost down the drain.
- Turning off your storage heater if you are going on holiday.

Water conservation when using appliances is looked at in the APPLIANCES chapter (p.94).

Priorities for action

1. If you heat all your water by electricity, consider choosing an alternative from the various options outlined in this chapter.
2. Using the criteria listed on p.81, decide whether an instantaneous or a storage system is more suited to your household.
3. If you are buying a new heating appliance then consider one emitting the least amount of CO_2 and consult **CHOICE OF ENERGY SOURCE** (p.80). A gas appliance is likely to be the most efficient choice in this respect. If you are combining your central heating with your water heating then read about gas appliances in the **SPACE HEATING** chapter (p70).
4. Undertake as many of the hot water conservation measures as you can. In particular:
   • Insulate your storage cylinder well. Remember it is losing heat all the time.
   • Use less hot water where possible either by shorter or less frequent use or by using constricting flow nozzles for showering or hand-washing.
5. Consider using a solar panel for your summer water-heating needs.

Lighting

We are witnessing a revolution in lighting energy efficiency: one that will make lighting a far smaller proportion of our energy requirements if we use these advances intelligently. Once electrical heating has been reduced to a minimum, lighting can account for anything from 30-70% of the total residential electrical consumption so lighting becomes the next priority to tackle, and it is encouraging to know that this is an area where savings can be made much more quickly and easily than in almost any other. So far in this chapter, save for the use of fans to aid ventilation and the control systems involved in heating, the use of electricity has generally been frowned on; however electricity performs this function with greater efficiency than any other means, except of course the use of direct light from the sun. Nevertheless, because we are dependent on this highly polluting energy source, we should use it with care and respect. There are a number of ways of reducing our energy demand as regards lighting:

- Using energy-efficient light fittings.
- Making the most of (sunlight and) daylight.
- Directing light for our needs (task-directed lighting).
- Controlling our use of light automatically.
- Maintaining the output from our lights by keeping them clean.

Energy efficient light sources

The majority of light fittings in most people's homes are tungsten incandescent bulbs. However 90—95% of the energy that is consumed by these fittings is converted to heat, with only 5—10% being emitted as light. How can such a basic inefficiency be reduced? There are two ways that light is emitted from the normal range of domestic light fittings:

- Incandescent light from a tungsten filament
- Fluorescent light from an excited phosphor coating

However, fluorescent light is four to five times more efficient than incandescent light! Both these systems of producing light have undergone recent technical improvements which we are only just beginning to make full use of domestically. They will now be examined in more detail.

Fluorescent lights

Fluorescent lights have a bad reputation. For many, the long tube that emits a bluish-white light that flickers and buzzes, reminds them of dull, unattractive environments. However many of the limitations of fluorescent lights are now being overcome.

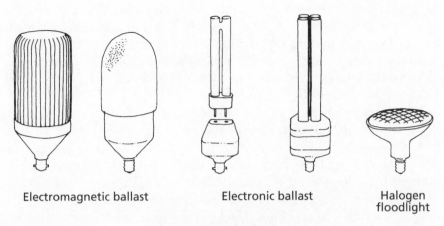

Electromagnetic ballast   Electronic ballast   Halogen
                                                             floodlight

Types of light fittings.

Fluorescent lights and tungsten filament bulbs work in entirely different ways. The fluorescent tube contains a special gas at low pressure. When an arc is struck between the lamp's electrodes, ultraviolet radiation is produced from the collision between the electrons in the arc and the atoms in the gas. This ultraviolet light impinges on a phosphor coating on the inside of the tube and excites it to produce light at visible wavelengths. The quality of light that is produced depends on the precise mix of phosphors in the coating: some produce a light more akin to daylight, while others mimic the tungsten lights, producing a more reddish-yellow colour. All fluorescent lights require a ballast to operate in addition to the tube. The ballast regulates the voltage and current and is essential for the lamp to operate properly. Each different model of lamp requires a different electronic input and hence requires a ballast specifically designed to drive it. The core-coil ballast, which has been the standard since fluorescent lighting was first developed, uses electromagnetic technology. The electronic ballast, only recently developed, uses solid-state technology.

Compact fluorescent lights (CFLs)

In the past fluorescent fittings were confined to long tubes. New technology has shrunk these tubes such that they are now like small cylindrical fingers protruding from a slightly enlarged bulb base. These compact fluorescent 'bulbs' are now being developed to replace every tungsten filament application. It is only a question of how long this process of substitution will take.

Although there may still be some uses for the old fluorescent tubes, it is the new compact fluorescent fittings with electronic ballasts which will be the main interest of home owners. These electronic ballasts have several advantages. Standard fluorescent lights flicker at 50 cycles per second, which has proved a problem for some people, producing headaches and nausea in certain situations. However, electronically ballasted fittings oscillate at more than 20,000 cycles per second, far faster than the eye can detect. They are also a third more efficient than the electromagnetic ballasts and start almost instantly. Dimming options are being developed for this type of bulb which will never be possible for the electromagnetic type.

These new lamps present us with a wonderful opportunity to light our homes with a quarter of the energy required before. Virtually all tungsten fittings should be replaced with compact fluorescent fittings as the opportunity arises. Although they cost more than traditional bulbs, you start making savings in your electricity bill straight away. These lamps fit into the same sockets as your normal bulbs and give you a warm coloured light similar to tungsten fittings; they are generally four to five times as efficient and last five to ten times as long. Install these first where they are used most—in your living areas and circulation areas that you like to have lit more continually. When cold they take a minute or so reach their full brightness. There are special bulbs for cold outside conditions.

Incandescent tungsten filament

Incandescence literally means light produced from heating. The heating is achieved by passing an electrical current through the thinnest strand or filament that can support itself. The filament is delicate and gradually blackens, eventually burning out. The electric light bulb is thus just one step removed from the gas mantel.

Quartz halogen lights

These are the latest souped-up models of the tungsten filament lamp. They produce a brighter, whiter light and are more energy efficient because they operate their tungsten filaments at higher temperatures. What is more, the blackening of the filament that occurs in standard bulbs, reducing their output by 25%, has been largely overcome and the light they produce decreases by less than 10% over the lifetime of the fitting.

All this has been achieved by enclosing the filament in a quartz envelope which withstands the heat better than glass, and this envelope has been filled with a halogen gas. When the lamp is operated, particles that evaporate from the filament combine with this gas and are re-deposited back on the filament when the lamp is turned off. This process both reduces the blackening and extends the life of the filament. Where halogen lamps are used with dimmers, these need to be turned on full from time to time to allow this regenerative process to take place. It is important never to touch the quartz envelope with your bare fingers as the natural oils in your hand will react with the quartz glass and cause it to fail. Often this envelope is enclosed in a larger bulb to prevent just such an occurrence.

Tungsten-halogen lamps produce about 50–100% more light per watt than standard tungsten filament bulbs and last longer. Depending on the model this could be from 2,000 to 4,000 hours.

Many models of quartz-halogen bulb operate at 12 volts which mean they require a transformer. These are now quite neat and small, so don't be put off by this. The main drawback of these lamps is their cost but their energy saving quality makes up for this. They do give off a very attractive sparkly light—their speciality is that the light source can be focused and directed better than any other light source. This makes them particularly appropriate and efficient for task lighting.

Long-life bulbs

These bulbs have been around for a long time and are at the moment being advertised as the latest thing. However they are very similar to existing bulbs except that they have a thicker filament and incorporate a special gas in the bulb which makes them last up to four times longer. The irony is that instead of using less energy they often use more to produce the same amount of light. They have their place perhaps in relatively inaccessible fittings but not in normal domestic circumstances. So don't buy these bulbs if you want to save energy.

Lifetime and efficiency comparisons

	Life time (hrs)	Efficiency (lumens per watt)
Standard incandescents	1,000	13
Longlife incandescents	3,000	12
Quartz-halogen	2,500	24
Compact fluorescents	7,500	60

Maximising our use of daylight

How can we make the very best use of sunlight and daylight—the one free source of light that is available to us? It is useful to remember that daylight, when entering through a standard-sized unimpeded window during the middle of the day, can amount to the equivalent of between 5 and 30 hundred watt bulbs.

Daylight and sunlight are very important factors in providing a healthy home that is not over-reliant on artificial light. The most important means of achieving this is through windows: but in existing housing we are limited by the size and position of existing windows. There follows a list of measures that we can take which show that there is often much we can do to affect daylight levels in any particular room in our home:

- Internally, use a light coloured paint or wallpaper a room to make the most of any daylight entering the room. Avoid dark carpets, dark-coloured furnishings or other dark floor coverings. Even the use of carefully placed mirrors, reflective louvres or blinds can bounce the sun's light deeper into a room.

- Externally, ensure that as much light as possible travels through our windows by painting the reveals and frames white and thinking of any other ways of reflecting light in, such as using reflective surfaces on a path outside or painting adjacent garden walls white. Even a pond will reflect light in through a window. You can also reduce any obstructions that might be overshadowing your windows: foliage might need cutting back, trellises removed, etc. Curtains can be kept away from windows by using longer tracks or curtain ties to hold them back. Don't forget to keep glass clean, as this can reduce light considerably—without our realising it.

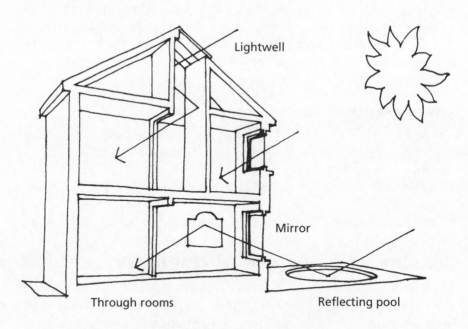

Optimising the use of daylight.

- Look at areas of your home that have little daylight or sunlight to see if there are any alterations that will improve this: you may consider increasing the height of windows if major renovations are being carried out, or adding rooflights. As a rule of thumb, if you can see the sky you are probably receiving sufficient daylight at the point where you are standing. Light wells through the roof are very effective if there is unused roof space above a poorly lit room. Rooflights admit considerably more light than vertical windows. Another radical alternative is to think of opening up the middle of the house and bringing a lightwell down the middle.

- Another way to make the most of daylight is to arrange rooms around the house to make use of the sun's path across the sky. In spring and autumn we can make the most of daylight hours by going to bed and getting up earlier. We can also choose to work or read wherever daylight is most available.
- Too much light can also be a problem by causing uncomfortable glare. You will need to weigh up the conflicting considerations of outlook, visual comfort and the admission of daylight. Glare can be moderated with venetian blinds or light curtains.

Directing light to where it is required

There are three different types of light that are referred to by lighting designers:

- General—e.g. a central hanging light
- Task lighting—e.g. a desk lamp
- Atmospheric or ornamental—e.g. a spotlight on a picture

Directing light where required.

The most important principle of energy-efficient lighting design is to make the most of any light sources available, and direct the light to where it is most required, either for a particular task or for effect. Often we will have a whole room lit to a reasonably bright level or even the whole house lit and yet we are working away with barely enough light for the task in hand, be it reading, sewing or writing. This is clearly an inefficient use of resources, yet many of us do it unconsciously. It is also possible to produce exciting or soothing environments determined by the fittings chosen and the way these are directed. Here are a few guidelines that can help with getting your lighting right.

- Think carefully which lights you really need or whether it is possible to reduce the wattage of any light fittings. Also, use bulbs appropriate to the fitting/fixture. In a recessed or spot fixture, a reflector lamp of half the wattage delivers the same amount of light obtained from an ordinary bulb.
- Use task lights. Decide what you really need the light for—then design your lighting scheme. Provide reading lamps where you need to see detail. Reduce your background lighting levels with lower-wattage bulbs to save energy and create more contrast in a room. The ultimate in task lighting is the miner's helmet that shines a light at whatever the head is pointed to. For some activities this might even be the most appropriate solution.
- Light for safety. Light outside steps with low level lights and do the same with inside stairs at night. Place a small light near the keyhole of an external door.
- Light for effect. Use dimmers with a halogen lamp as an energy-efficient way of varying lighting. Highlight the best features of your house and if you have special high-wattage decorative lights in your home then use them sparingly on special occasions. These are all ways that lighting can make your home more attractive. Light coming from multiple directions is also more attractive than lighting coming from only one

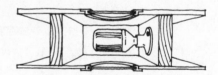

A light in a hole in the floor.

direction. Also have some fun with special effects, such as mirrors to multiply the images; or use a light in glazed hole in the floor to light both up and down at the same time—say on the landing of a stairway.

Controlling light

Occupancy sensors, photocells, and timers

As with all things electric, there is the possibility of all manner of controls that can make systems more or less efficient by providing light only when and where it is needed. This is mainly achieved by infra-red sensing of a person's presence. These systems have been developed and used for security systems, but are increasingly being used in commercial environments to save energy. Domestic applications will follow as the cost of these controls comes down. The commonest present domestic application is that of external lights activated by infra-red sensor.

It may be some time before all our lighting is on automatic and in the mean time we must use our little grey memory cells! However there are two things we can do to help:

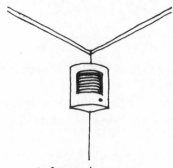

Infra-red sensor.

- Make sure every light has its own switch so that only lights that are needed are on.
- Put switches in convenient places to encourage the turning on and off of lights as often as is required.

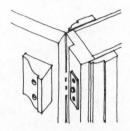

Cupboard door switch.

Priorities for action

1. Use energy-efficient lamps where possible, changing the bulbs in most used locations first.
2. Use lighting only when and where it is needed; this may require some careful thought and planning.
3. When setting up task lighting remember that you need only use a low wattage lamp if positioned close to where light is required.
4. Use tungsten halogen bulbs sparingly to add sparkle to selected locations.
5. Make the very most of your daylight potential by reflecting as much light as possible into your home, reducing obstructions.
6. Use controls such as timers, infra-red switches and contact switches for cupboard doors.
7. Keep all your bulbs and tubes dusted to maintain light output. Accumulated dirt has a significant effect.
8. Consider altering your waking day as far as possible to harmonise more closely to sunrise and sunset, particularly in spring and autumn.

Appliances

There is an ever-increasing profusion of modern appliances, most of which are sold and bought with very little consciousness of the energy that they will consume. Our kitchens particularly tend to be full of them. However this consciousness is changing, helped by the increasing number of green consumer publications. In some cases it is not so much a question of which appliance is the greenest choice, but whether such an appliance should be used in the first place.

Profusion of appliances.

The last chapter dealt with lighting, an easy area to make energy savings almost instantly by exchanging light bulbs. This chapter demonstrates opportunities to conserve energy by improving the efficiency of some of the rest of our most energy-exploiting appliances. In many cases we find we can simplify our lives at the same time. The following appliances, that assist us in the two main functions of feeding ourselves and maintaining our clothes, have been found to use the most energy:

- Refrigerators and freezers
- Cookers
- Washing machines
- Tumble dryers
- Dishwashers

Choosing an energy-efficient appliance

What are the key considerations you should have when thinking of buying one of these machines ?

- Do you really need it in the first place? As we become aware of how much damage to the environment one of these machines can do in its lifetime, we should seriously consider whether we can do without certain appliances such as dishwashers.
- The energy source—For many of these machines there is no choice but electricity, but where there are alternatives these should be used if at all possible.
- Durability and ease of repair and maintenance. These are both crucial to the life expectancy of a machine. Very often this is a low consideration for manufacturers compared to the initial cost and extra gimmicks.
- The energy in use—this is the most important consideration in the case of nearly all these machines, and one for which it should be possible to make comparisons before making a purchase. Manufacturers make a big play of the initial capital cost but much less of cost in use.
- The energy required to manufacture and dispose of the appliance— there are no comparative figures available yet, but this is clearly an area of increasing importance.

Lifecycle energy cost

The last two considerations can be brought together in the lifecycle energy cost:

MANUFACTURING ENERGY + OPERATING ENERGY + DISPOSAL ENERGY
= TOTAL LIFECYCLE ENERGY

More commonly this is expressed in monetary terms:

PURCHASE PRICE + LIFETIME ENERGY COST = LIFECYCLE COST

For example, if (a) you purchased a refrigerator for £250 and it costs £45 a year to run, or (b) you purchased a refrigerator for £350 and it costs £30 a year to run, and both machines have a life expectancy of 20 years then:

(a) £250 + (£45/yr. × 20 yrs.) = £1,150
(b)£350 + (£30/yr. × 20 yrs.) = £950

This simply demonstrates that running costs in terms of energy and money are often an equally important consideration as the initial cost when it comes to buying a machine.

From an ecological perspective it would be more useful to be able to express all the costs, including the manufacturing costs, in terms of CO_2 produced. This would be even more revealing than just comparing the capital cost with the running cost. If you go for an ecological choice and choose an appliance which will both last a long time and use minimal operating energy, you will almost certainly have to pay considerably more initially. However this extra could well be saved several times over during the life of the appliance. More importantly you will prevent many tons of CO_2 entering the atmosphere.

Reducing our energy consumption by careful use

How should we view the machines we already own? Remember that we have a large influence on how much energy any appliance consumes, as it is how often and how efficiently we use a machine that probably has the greatest influence on its energy use. Undoubtedly there are different ways reductions can be achieved:

- Through choices relating to the overall use we make of high energy consuming appliances. For instance, how much raw food we eat rather than cooked food. Are we willing to think of our meals in terms of the energy consumed in their preparation?
- By using the heat being generated in our appliances, particularly in winter. This might affect the way we use appliances at different times of year.
- By co-ordinating the amount of cutlery and crockery we use with the most energy-efficient wash cycle of our dishwasher (find out what the cold rinse will achieve).
- By becoming aware of energy wasting habits such as opening fridge and oven doors unnecessarily.

Refrigerating and freezing

In recent years, the efficiency of refrigerators has increased substantially. Manufacturers have:
- Improved compressors and motors
- Introduced better door seals and compartmentation
- Increased insulation standards

Even though there have been improvements, most fridges and freezers still operate a long way short of the maximum potential efficiency. Average running costs could be cut by half without serious difficulties, as we can see when we realise that the most efficient existing appliances are 50% more efficient than the average. Although a refrigerator uses a relatively small amount of power in use, it is the fact that it is working 24 hours a day 365 days a year, often in the hottest room of the house, that causes it to be one of the highest energy users.

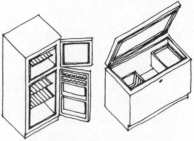

A chest-type freezer (right) loses less heat when opened.

Choosing a new energy-saving fridge or freezer:

- Unless you use a large amount of frozen food it makes the most sense to buy a combined fridge-freezer model. The most efficient place for the freezer compartment is at the top so that it can facilitate cooling below.
- If you do buy a separate freezer, then buy a well-insulated chest type which loses far less cold air when opened than an upright.
- Work out carefully the optimum size you need. Both fridges and freezers operate best when at least three-quarters full.
- Automatic defrost models consume far more energy than the manual defrost models. This is because they contain heaters. The fridge is thus working against itself when defrosting and if the fridge is being used inefficiently the energy used can be increased by a further 50%.
- Find a fridge that is well insulated and look for energy efficiency labels.

Energy saving in use

- Energy saving microelectronics have been developed for refrigerators: they help to match the real load on the motor and save up to 20% of energy. These electronics have been incorporated into a special plug called a saverplug which is well worth installing.
- Locate a refrigerator or freezer away from sources of heat such as a stove or direct sunlight. If you have a fridge that is not well insulated you can provide some extra insulation yourself on the top and sides.
- Check the temperature inside the main refrigerator compartment and freezer. They should be between 3° to 5°C and -18° to -15°C respectively. If the temperature is outside these ranges then energy is being wasted and the temperature control needs to be adjusted. Fridge and freezer thermometers are well worth buying and placing

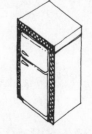

Extra insulation to fridge

in your fridge for the occasional check.
- The back of a fridge is a notoriously difficult place to clean. However the coils or fins tend to become sticky and dusty—this acts as an insulant forcing the motor to work harder and waste more energy. Clean with soap and water.
- If you have a manual or partial automatic defrost facility then defrost regularly.
- Test the door seal and check that it is tight. This is equivalent to draught-proofing your fridge. If there is an air leak it can not only lose energy unnecessarily but will require more frequent defrosting.
- Dispose of your fridge where the CFCs can be salvaged. Enquire at your local council to find out where you can take it.

Cooking

Food can be cooked on a gas or electric hob, in a gas or electric oven or in a grill (usually only electric). These different ways of cooking food can also be combined, as they usually are, in a stove. This stove in turn can be either all electric, all gas, or best of all a gas appliance with electric ignition and energy saving controls. There is yet another approach to cooking where space heating, water heating and cooking are all combined in an appliance such as an Aga—a logical integration descended from the old kitchen range. However, at present Agas lack the sophisticated electronic controls which could make them more energy efficient.

The initial cost of electric cookers is less than that of gas cookers. However, tests carried out by *Which?* magazine have shown that electric cookers cost about *three times as much to use* as gas cookers. If we then compare the amount of CO_2 produced it could amount to four times as much! There are some electric cookers that are more efficient at transferring the energy to the pot or the food, such as microwave ovens or induction loop cookers:
- Microwave ovens—these save up to a third of the energy when compared to a standard electric cooker, making them a little more competitive. However there are a number of factors involved in the energy usage, such as the amount of food that is being cooked and the receptacle that is being used to hold the food.
- Induction loop cooking—saves about 10-20% over conventional electric cookers. Such cookers are not very common.

These savings are insufficient to overcome the ecological disadvantage of using electricity for cooking.

Choosing an energy saving cooker
- In terms of the choice of fuel, gas is the obvious choice if you have a

mains supply (use propane or butane if not).
- Buy a gas stove that incorporates electronic controls to save energy and save on wasteful pilot lights.
- Cookers are not very efficient at cooking small quantities of food—plan carefully how you intend to do this at the same time as buying your stove and pans. A microwave may sometimes be a more efficient answer if you are heating a lot of small portions. Having a range of pans of different sizes that are shaped correctly for the type of stove you have is also important.

Two final points should be made here that have a bearing on your consumption patterns, as well as on energy conservation:
- Remember that good quality food is generally healthier if you eat it raw rather than cooked!
- Eat less! On average in Western society we eat far too much.

Clothes washing

Washing machines use energy in two ways: to power the motor and to heat the water. Heating the water accounts for some 90% of the energy consumed. Most of the machines sold today heat the water within the machine which means that the water is heated electrically—something to be avoided if possible. Therefore if you heat your hot water by gas it is worth buying a machine that can use hot water from this source even if it has an electric booster heater.

Features of energy-efficient washing machines

- Front loaders use about half the energy of top loaders as they use far less water.
- Machines that use gas-heated water are far more energy efficient.
- Machines with controls that allow the temperature and water level to be adjusted or have energy-saving programmes will use no more hot or warm water than necessary for any particular load.
- Machines with a faster spin save drying energy.

Energy saving in use

- Most laundry loads can be washed in cold or warm water. Use hot water only when essential.
- Put soap directly where it is required—on collars and cuffs. Soaking clothes ahead of washing can mean a shorter or cooler wash.
- Follow the instructions to obtain the most efficient use from your machine and aquaint yourself with each programme to understand

which are the most energy efficient. Always try and fill your machine to capacity.

Clothes are generally being washed more and more frequently due to the ease of using machines. It is worth remembering that in the eighteenth century some girls were sewn into their dresses, which were only undone for a bath at Christmas and Easter! I am not suggesting that we follow this practice, but the current habit of throwing all manner of clothes into the washing basket after wearing once is hardly sustainable.

Clothes drying

Using an electric tumble dryer is not an ecological way of drying your clothes. There are many alternative ways of drying clothes, such as outside on a clothes line or on a clothes rail above a heat source such as a central heating boiler. In tumble dryers, about 95% of the energy used by the machine is for heating. Gas is the most ecological way of heating, so if you think it is necessary for you to have a clothes dryer, a gas-powered one would certainly be preferable. Here is some advice on saving energy:

- A moisture control sensor to stop the machine running when the clothes are dry is essential. Over-drying wastes energy, shrinks clothes, shortens fabric life and generates static.
- Different heating levels to allow light drying will prevent clothes from becoming too hot and allow greater flexibility of use.
- A cool-down cycle will save energy by using the heat in the clothes and machine to continue to dry and prevent wrinkling.
- Timers are only effective for saving energy if it is known how long a particular load will take.
- A big drum is more efficient than a small one as it allows air to circulate properly. It is important to dry the right amount of clothes at any one time—both too many and too few will waste energy.
- Check the dryer exhaust vent to ensure it doesn't leak. The flapper on the outside should open and close freely. A flapper that remains open may also encourage unwanted draughts.

Dishwashing

There are much greater differences in the amount of energy that dishwashers use than with almost any other machine. This appliance is yet another example where heating is usually undertaken with electricity. If therefore we can use gas-heated water and also a programme that limits the water used, then using a dishwasher becomes more acceptable. It is obviously important to find out how much energy is used in different cycles; you may be able to find machines that only use hot water for a very short period of time. But really, in the last analysis, do you need one of these machines at all?

Features of an energy-efficient dishwasher:
- A machine that uses hot water from your cylinder is more energy efficient so long as your water isn't heated electrically.
- Other features which save energy are:
  - Short cycle selector
  - Air dry selector
- If buying a machine then get one that is the right size for your needs and holds enough crockery and cutlery so that you can always run a full load.

Priorities for action

1. Consider carefully the need for a machine that requires large amounts of energy to be heated by electricity. Be imaginative about the alternatives.
2. If you do decide that you need a particular appliance, think of ways that its use can be reduced to a minimum.
3. Find ways of harnessing your machines to non-electric sources of heating.
4. Identify machines that have the most energy-efficient cycles.
5. When buying an appliance, aim to buy one that is durable, easy to service and has the lowest energy consumption, so that its life cycle cost is reduced to a minimum.
6. Use 'green consumer' publications to find out the latest information on energy saving appliances.

Solar energy

The sun is the source of all energy on earth except for geothermal and nuclear energy. Even fossil fuels are the product of ancient solar energy. We can feel this energy very directly when we sunbathe and are warmed by the sun. We are also using this direct energy all the time in our homes as it warms the fabric and as it enters through windows and becomes trapped by the greenhouse effect (in its original meaning). This use of solar energy is called passive solar energy as no mechanism is used to enhance or collect it. Active solar energy uses special collectors, of which there are two main types: fluid collectors which heat a fluid circulated within them, and photovoltaic cells which convert light energy to electrical

Solar power.

energy. Both these active ways of harnessing solar energy are dealt with first.

Active solar energy

Solar water heaters

There have been many different types of solar water heater installed in Britain. However very few of the early models worked with any degree of satisfaction, and many members of the public have been sold systems that cost hundreds and sometimes thousands of pounds which resulted in no appreciable reduction in their overall energy expenditure. In Britain it is only worth using the most advanced and efficient systems to take particular advan-

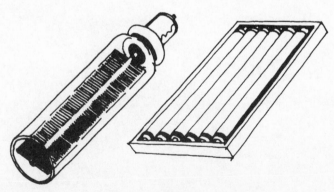

Solar collectors.

tage of the summer sun to provide hot or warm water, so that the main boiler can be turned off. The far south of the country will provide considerably better returns than the far north. Where the total amount of heat collected is small, the most successful method of using this heat is to preheat the water in the hot water cylinder, so that less energy is expended to bring it up to the required level.

The way the most advanced existing solar collectors work is to collect the heat in vacuum tubes similar to those used in fluorescent light fittings. This reduces the heat loss considerably. In some of these heaters the back of the tube has been silvered so that light and heat are focused on to another long thin blackened tube containing water or oil at the centre. Others simply use an absorbing metal plate with the liquid similarly running down a tube in the middle. This second type is more suitable for overcast skys and so is most commonly used in Britain. The heat in the inner pipe is transported away by a liquid medium to a calorifier or heat exchanger, usually inside a hot water cylinder or in a pre-heating cylinder.

A second type of collector, which is more commonly used, is the flat plate collector, consisting of a metal absorber plate to which tubes containing the heat transfer fluid are bonded. The absorber plate is placed on a layer of insulation and the whole is put in a box with a glass or plastic lid.

If you have a large enough area of these collectors it is possible to obtain all your hot water needs during the summer and a considerable proportion of your autumn and spring needs. This will depend on your having a suitable roof, or equivalent inclined location, facing roughly south with enough area to take the number of collectors you require for your needs. The main problem is that to install a really efficient system is still relatively expensive. Generally, the most energy-efficient plan is to complete all your energy conserving measures first before investing in solar water heating. However, if electricity is your only means of heating, then fitting solar panels makes sense.

Orientation and mounting of collectors

The optimum panel size for an average household is about 4 square metres: a larger panel area will of course provide more hot water but not so cost-effectively. It is important when mounting a collector to watch for obstructions such as buildings or trees; the panel should ideally face no more than 45 degrees from due south, i.e. south-west or south-east, but not beyond. For optimum solar collection over the whole year it should be tilted at an angle of about 35 degrees to the vertical. Care should be taken with the mounting of the panel so that the weight is taken by the rafters. A low power pump is used to transfer the heat from the panel to the hot water cylinder and a differential temperature controller is used to turn the pump on when sufficient solar heat is available to heat the tank. It is important that this is properly installed, as otherwise much energy can be wasted. Neither building regulations consent nor planning permission is usually required but it is worth a phone call to check with your local authority. However, in a conservation area, planning permission is likely to be needed.

Photovoltaic cells

Photoelectric cells are based on a wafer or ribbon of semiconductor material, usually silicon. One side of the semiconductor material is electrically positive (+) and the other side is negative (-). When light strikes the positive side of the solar cell, the negative electrons are activated and produce an electric current. When a group of solar cells are connected, a solar module is created which then produces a higher voltage than just a single cell. The cells are encapsulated under transparent material and the total electrical output is determined by the number of cells that are connected together within a module. Modules can be further linked in a panel to form a solar array.

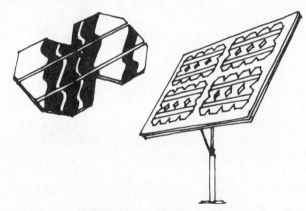

A photovoltaic cell and array.

The efficiency of most standard crystalline silicon cells is about 13%, which means that 13% of the energy in the incident light is converted into electricity. A photovoltaic device converts light into DC (direct current) electricity. They have become common in powering calculators and watches that require very small amounts of current and have proven to be reliable electrical power producers in thousands of terrestrial and space requirements. There are continuing developments in photovoltaics with many possibilities for future developments in domestic energy applications. It is conceivable that windows themselves, or even south-facing walls and roofs, could become solar cells.

For the present, however, the same can be said of photovoltaic cells as for solar water heating—that for the equivalent expense you can probably save far more energy by undertaking other energy conserving measures first. Their expense precludes them from being a first priority installation in most people's homes. Their use at present is particularly appropriate in isolated situations where small amounts of power are required.

Passive solar power

The principle of passive solar power is very simple in its commonest form. Light energy enters through glass, which is transparent to visible wavelengths (as we know, because we see through it). This light energy is transformed into heat when it is incident on the interior of the room. When this heat is re-radiated from the surfaces within the room its wavelengths are longer and the glass, acting like a mirror to these longer wavelengths, reflects most of this heat back into the room.

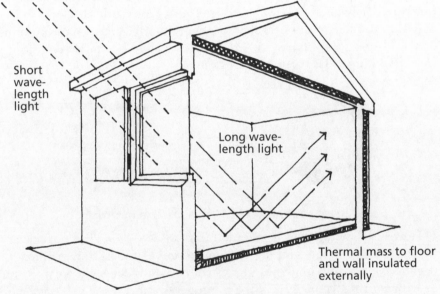

Short wave-length light

Long wave-length light

Thermal mass to floor and wall insulated externally

Passive solar gain.

If we want to make the most of this heat and keep it from dissipating too quickly, there are four main things we can do:

- Insulate the external walls, ceiling (or roof) and floor of the building in which we wish to contain the heat (see **INSULATING** chapter).
- Install double-glazing and use low emissivity glass to reflect even more of the heat back into the room.
- Increase thermal capacity if necessary: if your home is not constructed of heavy materials, such as bricks or concrete, that have a high thermal capacity to store and even out the heat, then it may be worth introducing some thermal mass within the insulated envelope (columns of water contained in glass have been used effectively).

- Cover the windows at night with insulating shades or shutters to reduce heat loss (see pp.66-67).

Any window can thus become a solar collector—even north-facing ones can become collectors during daylight if the insulation levels are high enough in the window through which the light passes.

It is important to realise that in Britain we are placed further north than we often perceive ourselves to be, because of the warmth of the Gulf Stream. The majority of the country is between 51 degrees and 58 degrees, which puts us on the same latitude as Denmark, Canada and mid-Russia. The USA, where solar energy is more used and more research done, is much further south than in Britain. Our comparatively northerly latitude means that we receive very little energy from the sun in the middle of winter, especially in northern Britain. November, December, January and the first half of February are all low solar energy months. These are the months when draught-proofing and insulation are so important, and when we almost certainly have to use top-up heating to keep our homes warm.

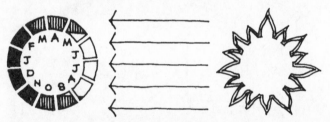

Incident solar energy through the months.

However if we think of the months of spring (the second half of February, March and April) and autumn (September and October), this is when we may be able to avoid using heating or reduce it considerably if we make the most of passive solar energy during this time. Another name that is given to the light energy that enters our houses, straight from the sun into our living rooms, is 'direct gain'.

Windows and rooflights

Windows are the most important element on the facade of a house. They give a house character. They are like the eyes through which we can look out and, as we have seen, it is this same property that allows the sun's energy to pass through the glass and into the house. Windows that face south, south-east or south-west receive the most sunlight throughout the year and even more is collected by an inclined south-facing rooflight or skylight.

Dormer windows provide another opportunity for passive solar gain. Between rooflight and standard windows, these often have the advantage that they are high up and miss the possible shading from buildings and trees. In closely built-up urban areas it is often difficult to achieve sufficient solar gain from downstairs windows. Besides this, solar energy falls at its most intense on an inclined roof. In most houses this energy remains unused and it is worth considering ways of using it. Besides solar collectors there is the possibility of creating a roofgarden, greenhouse, conservatory or sunroom. Some of these will require considerable redesign to incorporate them and of course this will not be possible for some houses. However your roof remains a possible area for development—and to have fun with.

Windows, rooflights and overshading.

For most houses in Britain, it is questionable whether it is worth extending windows or building new ones on the main facade of the house with the sole purpose of collecting solar energy, even on the south side. However since rooflights are relatively cheap, they may present you with your best opportunity to improve the contribution that passive solar gain makes to the heating of your home.

Night-time insulation

The importance of insulating windows at night has already been mentioned. This subject is dealt with in the **INSULATING** chapter (see p.66).

Solar room addition

If you are are thinking of extending the accommodation in your house on the southern side, you would do well to consider this taking the form of a solar room. We have already mentioned the use of your roof space for this purpose as it may be the only place available to you with enough sun. If you are serious about collecting and storing heat in this way, then you need to think about how the heat can be transferred to other parts of the house. Thermal storage is probably difficult to achieve in your roof space, as it will very likely suffer from a general lack of thermal capacity. It may be worth thinking of dissipating some of the excess heat with a length of ducting and a fan, and depositing it elsewhere in the house where it is actually needed.

A solar room can be made by converting or extending an existing porch, by carefully building an addition on to the side of your house (see **EXTENSIONS** chapter in Part One) or by glazing over part of your roof. The type of solar room that is best for you depends on just what you plan to use it for. There are two basic ways to use one:

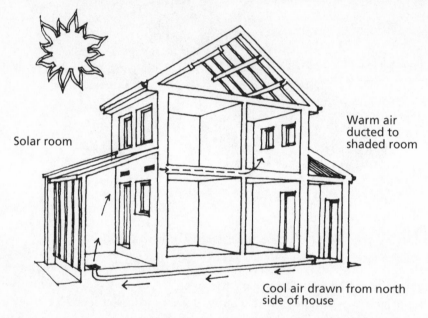

Solar room

Warm air ducted to shaded room

Cool air drawn from north side of house

Solar room used to warm cooler parts of house.

1. As a sunspace

Your solar room can simply be a warm, sunny place to sit during the day or a playspace for children. If you are at home in the day time, you can turn down the thermostat in the main house to save energy when is warm enough to use. If your sunspace is connected to the house, the heat can be channelled into the main part of the house and closed off again when the sun room cools off. You can wire a fan to the thermostat so that it turns on automatically when the temperature in the solar room reaches a certain level.

A more sophisticated way of doing this is to cut two vents in the wall separating the solar room from the rest of the house, one high on the wall near the ceiling and one down near the floor. You set a fan in the top vent to blow heated air into the house. Cooler air will return to the solar room naturally through the lower vent. Circulating solar heated air through the house reduces the need for heating and shortens the heating season. Best of all is to install a duct to carry heated air to the north side of the house. Unless this is done the room will almost certainly be too hot at times.

2. As a growing space

If the primary purpose of having a solar room is to produce food, flowers or air- or water-cleaning plants, you should think of it as being a solar greenhouse. You need to make sure that it can retain enough of the heat collected during the day so that the temperature doesn't fall below freezing at night. This can be done by adding thermal mass to store up heat or by using night-time insulation over the glass to reduce heat loss. What makes a greenhouse a solar greenhouse is that (with proper design) it takes little or no supplemental heat to keep it above freezing throughout the winter.

For many people the fuel savings from the attached solar room are less important than the amenity value and increased value of the home. The basic idea is that a good design will give you a room that doesn't cost you anything to heat. A *very* good design and proper distribution of heat will actually reduce your home heating costs.

Priorities for action

1. Consider the use of solar collectors for water heating. There are two main circumstances in which it makes sense: (a) if you have fulfilled the most important energy conserving measures and are looking to take further steps; or (b) you have an inefficient method of heating your water in summer, such as by electricity (see **WATER HEATING**, and **FLUID SOLAR WATER HEATERS** at the beginning of this chapter).
2. Maximise the potential your house has for collecting passive solar energy. This can be done by: (a) making the best use of your existing windows, and the best use of rooms that receive the most sun, particularly in the spring and autumn (see p.106); and (b) ensuring that as little as possible of this energy escapes at night, by insulating your windows with fitting curtains, blinds, shutters or pop-in panels (see p.67).
3. Harness solar energy by adding a solar room or conservatory in your roof space, or by building in an oriel window upstairs. Alternatively, you can build on a conservatory or sunroom. If you carry out one of these projects, it is important to be clear about its main purpose—whether for growing plants or for supplementary heating—so that it can be designed accordingly.

Part Three

HEALTH

Part Three

HEALTH

IT IS ONLY RELATIVELY RECENTLY that the broad ecology movement has begun to take seriously the threat to personal health from toxins, and the effects of various forms of non-nuclear radiation and even noise within the home. Since the last century the main domestic concern has been with cleanliness, space and light. Now, however, an environment that looks bright and clean may have heavily polluted air. The very materials that were designed to improve the quality of our lives now threaten this quality in other ways. This is most likely to be true of a new house where there has been heavy use of synthetic chemicals in timber treatment, plastics and petrochemical paints, but it can equally well be true of a newly renovated house. In extreme cases inhabitants become sick. What is more common, however, is a general increase in the number of complaints people have relating to allergic reactions and problems with their immune system. The air we breathe, the water we drink and the various forms of radiant energy that impinge on our bodies are all affected by the design and materials used in our homes and can have a significant effect on our individual health. In most houses today, contrary to what we may think, the quality of the air inside is considerably worse than that outside.

How can we deal with this complex ecological imbalance in our homes? Firstly, we need to identify the most serious threats and find ways of eliminating these as far as possible. We can then choose to go further if we wish and positively create an environment for ourselves that is both healing and life-enhancing. If we have these aims we need to inform ourselves of the basics; for it is only with understanding that we will be able to diagnose problems correctly and implement the changes that are necessary.

It is to provide such a basic understanding that these HEALTH chapters have been written. This is not a definitive tour of the subject, as much is still in the process of being researched. What I have endeavoured to do is to summarise some of the more important aspects of present knowledge. The first chapter deals with toxins and pollutants, their effect on our bodies and how we can avoid absorbing them. Chapters on air and water follow, where pollution is also the main issue and has a most direct effect on our health. Next come radiation and sound, both forms of wave energy which have an important bearing on health in the environment. Finally there is a chapter on plants, which surprisingly provide solutions to many of these problems.

But why should we be so concerned with our health? Is this not self-indulgent, or of minor relevance to the social and ecological problems with which we are faced? The following points make clear just how central our own health is:

- Our vitality can be one of the best indicators of the health of our immediate environment, our local ecosystem and our planet. If we can solve the problems of our own health we are almost certainly beginning to solve some of the worst ecological problems in our local ecosystem.
- Diseases with a strong environmental element involve enormous cost to ourselves and ultimately the planet. If the money spent on treating these diseases could be diverted to measures promoting positive health, by reducing the sources of toxins, we would achieve something far more positive.
- Finally, it is people healthy in body and spirit who are most likely to have the energy, enthusiasm and vision to solve the immense problems facing life on this planet. It is empowering to be in control of our own health and it helps us to take control in other areas of our lives.

It should be noted that this book does not deal with the dangers of fire, structural collapse or of accidents such as electric shock due to poor electrical installation. These are largely accounted for in existing building regulations and other traditional safety measures. Housing as built in its existing form in the West is regarded as 'safe' in a conventional sense. This part of the book will give you information to enable you to bring your own home environment up to a standard of positive well-being for the benefit of both you and the planet.

Toxins and pollution

We are hardly aware of the dramatic increase in the number of synthetic chemicals that are continually being invented and introduced into our lives. The chemical revolution, which at first seemed to herald such promising rewards, has been little controlled and there are now thousands of unnecessary and damaging synthetic substances loose in the biosphere. Perhaps something like 60,000 chemicals are in use around the world today. The effects of the more pernicious are becoming ever clearer in the atmosphere, the seas, the land and of course in plants and animals. Our health is under threat from these chemicals; some of them have infiltrated our homes unknowingly and others we welcome as solutions to domestic problems. They enter our homes either through the front door in the things we buy or through our water supply or ventilation systems. Sometimes they are in the very construction materials from which our homes are made. Not all these chemicals are synthetic—some are natural toxins, such as the radioactive gas radon which, in some parts of Britain, can infiltrate through the floor from the ground below.

What role can we play in turning this tide around before it triggers even worse effects? First we need to become knowledgeable about the sources of these toxins and stop, as far as possible, importing the most harmful offenders into our homes. It is useful if we can also understand the ways in which these toxins affect our bodies and the measures we can take in defence. We need to deal with the toxins that are potential threats in our existing homes. Finally, and perhaps most importantly, we need to know enough not to export toxins and pollution indiscriminately, such as throwing heavy metal batteries in the dustbin or flushing DIY solvents down the drain out of our homes. It is true that the biosphere can cope with amazing amounts of waste and pollution, but the overall quantities that it has been required to digest have been so great in recent years that there is a danger that our habitat is being irreversibly damaged. The more pollution we all export from our houses the more likely it will eventually return in our food, water and air. However, first of all we need to identify the most dangerous toxins and their sources.

Sources of indoor toxins and pollutants

Toxins in our homes originate from many sources and enter our bodies as gases, vapours, solid particles, liquids or radiation. They may be inhaled, ingested or absorbed through the skin. The following is a list of sources:

- Products of combustion—tobacco smoke, products of burning gas, coal, wood, paraffin, candles and joss sticks

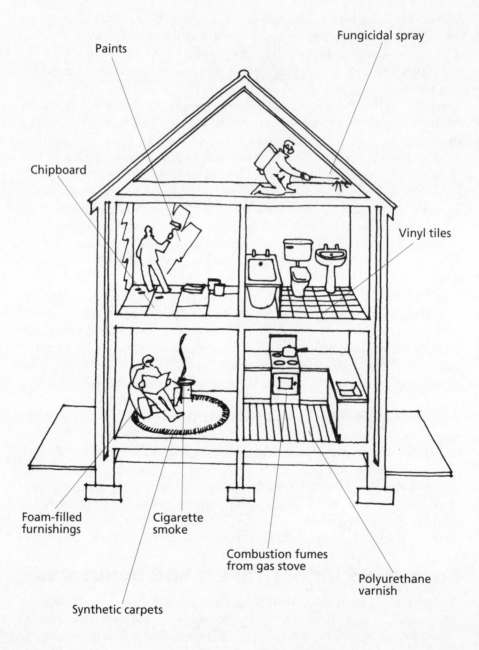

Paints

Fungicidal spray

Chipboard

Vinyl tiles

Foam-filled
furnishings

Cigarette
smoke

Synthetic carpets

Combustion fumes
from gas stove

Polyurethane
varnish

Sources of toxins.

- Vapours from household chemicals, solvents and paints
- Off-gassing★ from plastics, resins and rubbers, petrochemical paints and varnishes, insulation foams (polyurethane and polystyrene), plastic flooring, synthetic textiles in carpets, curtains and furniture
- Formaldehyde vapour from urea formaldehyde in glues, plywood, particle board (e.g. chipboard) and insulation
- Timber treatments against rot and infestations which can result in toxic dust and vapours
- Airborne fibre particles from asbestos and man-made mineral fibre insulation
- Pesticides used to spray indoor plants and insects
- Airborne particles of house dust, pollen from certain plants, fungal spores and droppings from dust mites
- Airborne pollutants from outside the home, such as car exhaust or local industry
- Water-borne pollutants in tap water, such as lead, nitrates and chlorine (see **WATER** chapter p.130)
- Radioactive radon gas from the ground & building materials (see **RADIATION** chapter p.140)
- Heavy metals in paints released through paint stripping
- Food, which contains a large range of chemicals including preservatives, pesticides and aluminium from cookware

There will be big variations in the extent of contamination in existing housing, depending on geographical location, how new a house is, the type of construction, and what finishes have been used. It will also depend on how often you redecorate and clean your home and the types of paints and chemicals used.

The effect of toxins on our bodies

Once a toxin is imported into the house, it can be difficult to stop it entering our body. If airborne it can be inhaled and enter the bloodstream through the lungs. If it is in contact with the skin it can be absorbed and pass into blood vessels directly below the surface. This can happen with solid materials such as lead (whose salts are dissolved and absorbed by moisture and oils in our skin), with liquid toxins, or with toxins dissolved in liquids which pass through our skin with even greater ease. In the case of many volatile solvents, such as petrol or paint solvents, there is danger from both absorption through the lungs (from the fumes), and direct absorption through the skin (from spills). Solvents, by their very nature, are absorbed very easily. Of course any

★ the slow release of vapours into the environment

toxins in food or drink that are ingested have an even more direct route to the bloodstream. These might include chemicals such as chlorine compounds in water or pesticide residues in food.

Once in the bloodstream these chemicals can be carried to every part of our body and can effect every organ. The nervous system is particularly vulnerable, since nerve messages are transmitted by chemical/electrical reactions and certain chemicals can have an effect out of all proportion to their concentration. Pesticides, for instance, are often specially designed to attack this system or to accumulate in fatty tissue associated with nerve cells. The more serious effects result in headaches, weakness or trembling. The kidneys and liver, as the organs responsible for processing and cleaning toxins out of the body, are vulnerable to abnormal quantities of almost any toxin, especially heavy metals, and the chromosomes in cells are damaged by certain chemicals which have the potential to set off cancer or cause genetic damage. What can make things more complicated and difficult to predict is that some apparently harmless chemicals can enter our bodies and react with other chemicals in the gut or bloodstream, creating unforeseen hazards.

We don't know yet how these chemicals will affect our long term health. However, an increasing number of people are experiencing allergic reactions, such as asthma and diseases such as Myalgic Encephalomyelitis (ME) and multiple chemical sensitivities (MCS), and most of us may feel 'under the weather' quite inexplicably more often than we feel we should. It is difficult to establish which are the most serious factors in our environment since many of the medicinal drugs we take, and stress-induced toxins from our own bodies, contribute in ways that make it difficult to separate one cause from another. However what is clear is that, for many people, the overall loads that their immune systems are expected to deal with are becoming too great for them to handle.

Our immune system and how we can support it

A bleak picture has been painted in the above section. Why haven't we all succumbed to the effects of accumulated poisoning ? Because we have in our bodies a most elaborate and sophisticated defence system against most toxins. Our environment has always had toxic compounds as part of its make-up, and life in all its forms has always needed to evolve ways of overcoming the poisoning effects of these. Human beings are no exception. Poisons can be seen as chemicals to which our bodies have not yet found an immune response. For example, plants produce prodigious numbers of chemicals, some beneficial, some neutral and some harmful. Our bodies deal with these as a matter of course through our immune system which is constantly working to keep our bodies healthy and in balance. In the past few decades the avalanche of chemicals and the general stress of modern Western life-styles, including the use of medicinal drugs, has put our immune system under immense strain. Changes in our environment that previously would have taken generations now can be measured in years or even months.

Apart from reducing the toxic stress on our bodies there is much we can do to support our immune system:

- We can eat a healthy diet and ensure we obtain the particular vitamins and minerals that support the immune system. We can also choose to undergo a process of detoxification (see references).
- We can learn to reduce stress levels as far as possible. For some it will mean learning to relax, for others reducing complexity and uncertainty in their lives and for nearly all of us, learning to deal as creatively as possible with whatever life brings.
- We can exercise, which stimulates blood flow, which in turn helps to flush out toxins. If exercise is difficult for some reason then massage can fulfil some of the same functions.
- We can also take enough fresh air and sunshine.

Although the above has nothing to do with renovating your home it has been included to show that our personal health and that of our environment are complementary.

Dealing with domestic toxins

Learning to identify the most dangerous toxins is a difficult task as there are many areas of uncertainty. We have a choice as to whether we do this in the more general form of knowing the likely sources of toxins in our homes (see list on p.117) or we can undertake a more specific approach by learning the categories of dangerous chemicals. A chemical classification of toxins is given at the end of the book and there are references for those who want to take the subject further. The approach you choose depends on the degree to which you wish to pursue the matter.

How can you deal with this problem in a practical way? You could list the chemicals and materials in your home that could be a health hazard, and estimate roughly how serious the hazard of each is likely to be. You then have a list that you can plan round. Your own level of health or those in your household is a critical factor in deciding how urgently you need to take action. If your general level of health is definitely below par it is worth acting quickly to reduce your toxic load. It is also possible to have tests to see whether there are materials to which you are particularly allergic. If however your health is reasonable, you might want simply to change your habits for the future, and accept that the existing toxic products you use will gradually be replaced with benign alternatives. It is up to you to set your own priorities.

Storing safely

There may be certain DIY and cleaning chemicals you wish to keep at home, but you will want to avoid any danger of their fumes polluting the inside of your house. There are two ways you can do this: you can store them in a well-ventilated outside store such as a garage, or you can create an inside store with sealed cupboard doors and ventilated from the outside via air bricks or a special vent.

Disposal

If you want to dispose of toxic chemicals, you find yourself in the same situation as many industries who have the same problem, but on a much larger scale. Putting them down the drain or into the nearest river is not a responsible solution. Nor can you simply throw them into your dustbin. It is necessary to find out who in your local council is responsible for toxic waste disposal. Some chemicals can be neutralised, some need to be incinerated in a furnace and others are biodegradable enough to be buried. There may even be ways of recycling some of them. If you are not satisfied with the advice of your council, try contacting your local Friends of the Earth group.

Priorities for action

1. Identify likely sources of toxins (see p.117) in your home. Making a list can be helpful, including an action column (see also CLASSIFICATION OF TOXINS and REFERENCES).
2. Toxins can be absorbed through the lungs (air), stomach (food and water) and skin (mainly through water, clothes and hand contact). Air and water are dealt with in the following chapters.
3. Toxins from food and clothes are beyond the scope of this book. However in general, the less processed and more organically grown food is, the healthier and more toxic-free it is likely to be. Almost the same can be said for clothes and bedclothes where natural fibres without toxic chemical treatments are the healthiest choice.
4. Skin absorption occurs with low toxic synthetics that you are in contact with for long periods of time (such as clothes) or highly toxic products that you handle for short periods of time, such as newly treated timber or lead. With the latter wear protective gloves and mask which are impervious to the toxic material you are handling.
5. If you are prone to symptoms that are likely to be produced by toxins, then apart from reducing the number of sources of toxin in your home, you can ensure you get enough of the necessary vitamins and minerals that support your immune system (besides any other measures that you take).
6. Dispose of toxic materials responsibly. Find out from your local council how they suggest dealing with any particular chemical or material. If you are not satisfied, get advice from your local Friends of the Earth group.
7. When buying chemicals such as household cleaners and solvents, look for biodegradability, and natural rather than synthetic sources.

Air

The purpose of this chapter is to look at the constituents of air, the pollutants it carries, and other relevant aspects that have a bearing on indoor air quality, including the positive measures we can take to improve air quality in our homes.

Air and water are intertwined in their role as the life blood of the planet, being the prime agents of cycling and recycling the elements of life. There are an enormous number of different elements and compounds that are cycled through the biosphere within this circulation system. Many of these cycles are only just beginning to be understood. We humans are now using this system for our own waste disposal purposes, assuming that it will recycle and absorb anything we throw into it. On a global scale we are beginning to see just how false this assumption is. On a domestic scale we have problems of a different nature.

Quite sensibly we are reacting to the need to conserve energy—in part to control global warming—by draught-proofing our homes. Without adequate ventilation, we are in danger of concentrating our own home-produced pollutants to the point where our health may be threatened. Internal air quality is undoubtedly one of the key factors in the make-up of a healthy indoor environment. Most of us spend an average of 90% of our lives indoors, the highest proportion of this being in winter, when indoor air pollution is at its worst. We are thus much more likely to breathe in and absorb into our bodies any gas, vapour or airborne particle that escapes into the air indoors.

Air composition and its pollutants

The following gases are all constituents of air, and have been identified for their relevance to indoor air quality. Their normal atmospheric percentage by volume is given in brackets (but indoors the relative proportions may vary considerably).

Nitrogen (N)

Nitrogen (78%), the most abundant gas, plays an important part both as a sustainer of atmospheric pressure and as a natural dilutant of oxygen (see below). Nitrogen is also an essential element for life: it is contained in proteins and DNA, and is extracted from the air by nitrogen-fixing micro-organisms in the soil.

Oxygen (O)

Oxygen (21%) is the second most abundant component of air, and is both the most vital and one of the most destructive elements in the atmosphere. It is highly reactive—evidence of this is seen every time we witness a fire. It is likely that fire itself is the reason for the percentage of oxygen in the atmosphere remaining stable at 21%. If it is higher than this, the likelihood of spontaneous combustion climbs rapidly (a 1% rise doubles the likelihood of combustion and at lower than 15% it does not take place.) Oxygen performs metabolic combustion in our muscles, providing motive power for our bodies. As an oxidant it kills bacteria and thus acts as an antiseptic. However its toxicity also has an impact on the ageing process and may be a major factor in setting our life span.

Carbon dioxide (CO_2)

Carbon dioxide (0.032%) is a key metabolic gas, influencing climate, plant growth and oxygen production. For such an important gas it is surprising that it is such a tiny proportion of air! In terms of its role inside the house, CO_2 most usually occurs at slightly higher concentrations than outside as a result of exhaling by inhabitants and flueless combustion. Running flueless equipment in poorly ventilated space leads to the build up of the products of combustion. CO_2, although the most abundant product of combustion, is the least harmful of the gases produced. In high concentrations it leads to a feeling of drowsiness and stuffiness. (The other main products of combustion, carbon monoxide, nitrogen dioxide and sulphur dioxide, are much more dangerous).

Methane

Methane (0.00015%) is the major constituent of natural gas; its role as an indoor air contaminant is in most cases negligible. The exception is where methane seeps from the ground or leaks from gas pipes, as it can build up in unventilated spaces and cause an explosion.

Carbon monoxide

This is one of the more important of indoor contaminants because of its extreme toxicity. It is produced by incomplete combustion. Within buildings it may result from tobacco smoking or flueless cooking or heating appliances. Breathing in CO causes hypoxia: the combining of CO with haemoglobin in the blood, cutting off its ability to carry oxygen, results in tissue injury and even death. It used to be a constituent of town gas before natural gas. Indoor concentrations rarely reach serious levels; however injury and sometimes death has resulted from the back-flow of combustion products from incorrectly installed water heaters.

Sulphur dioxide (SO₂)

Sulphur dioxide, a pungent gas and one of the gases of combustion produced from burning coal, wood, oil or paraffin, was once responsible for urban smog and is now the main cause of acid rain. It is rarely a problem now inside the home, except when produced by combustion with inadequate ventilation.

Hydrogen sulphide (H₂S)

Hydrogen sulphide is a very poisonous and very smelly gas (rotten eggs) produced by protein decomposition and burning sulphurous coal. You would normally be repulsed by its smell long before it became dangerous.

Ammonia (NH₃)

Ammonia is an important gas in terms of ecological cycles. Cleaning fluids are the most common source for indoor air pollution. As it has a strong smell, it is possible to avoid dangerous levels if you remain aware that strong, unpleasant smells are almost certainly unhealthy for you.

Environmental tobacco smoke

Environmental Tobacco Smoke (ETS) is a mixture of exhaled mainstream smoke (MS) and the side-stream smoke (SS) which evolves between puffs. ETS comprises a mixture of several thousand constituents, some of the more commonly measured of which are: particulates, nicotine, carbon monoxide, benzene, nitrogen dioxide, acrolein and formaldehyde. Besides the objectionable smell, high concentrations cause irritation to the eyes, nose and throat, as well as a reduction in visibility. Irritation can be further exacerbated by low humidities. Long term effects are an increased incidence of respiratory disease amongst children of smoking parents and an increased risk of lung cancer. The dangers of passive smoking are at last being realised and the banning of smoking in public places is gathering pace. It is also an issue each household needs to confront.

Organic compounds

As regards the air quality in buildings, a growing problem has been the increase in the number of volatile organic vapours. Some examples of these, along with their sources, are given below: little is known of the effects of exposure to low concentrations of many of these substances, particularly the effects

of exposure to mixtures of them. However if you have a general suscepti-
bility to indoor pollution it would be wise to reduce your exposure to these
chemicals to a minimum.

Volatile organic vapours	Commonly found in:
Acetone	Nail varnish, tobacco smoke
Benzene	Adhesives, spot cleaners, paint remover, particle board
2-Butanone	Caulking, particle board, floor coverings, fibreboard, tobacco smoke
Carbon tetrachloride	Grease cleaners, dry cleaning fluid
Chlorobenzene	Paint solvent, DDT, phenol
Ethyl benzene	Floor and wall coverings
Methylene chloride	Paint removers, aerosol finishers
Styrene	Insulation foam, fibreboard, tobacco smoke
Toluene	Adhesives, sealing tape, some wallpaper, paint, plastic floor covering
Trichloroethane	Cleaning fluid, correction fluids
Trichloroethylene	Paint solvents
Xylene	Adhesives, wallpaper, flooring
Ethyl acetate	Linoleum, varnishes, perfumes, artificial leather

(source: Buildings and Health—The Rosehaugh Guide)

Formaldehyde

This is a vapour given off from urea-formaldehyde glues, foams and certain
plastics. As a result it 'off-gasses' from particle boards that use formaldehyde
glues, such as chipboard and insulating foams. This can be a particular problem
where new building work has been carried out or new chipboard-based
fittings or furniture has been installed. This vapour could build up where
there is poor ventilation.

Wood preservatives

These chemicals are a mixture of solvents and pesticides such as
pentachlorophenol, gamma-HCH or Lindane, dieldrin and tributyltinoxide
(TBTO). They are extremely dangerous. Exposure to the active ingredients
in treated buildings is usually through the inhalation of dust particles which
have the compound attached to them. As well as being irritants and nerve
poisons, some of these compounds target specific organs, such as the liver,
and can accumulate there from a number of sources—including through the
food chain from agricultural use. Dieldrin is the most dangerous in the longer
term—for householders the greatest risk is from DIY work without full

protection. Newly treated timber should be left well ventilated for as long as possible.

Asbestos fibres

The name asbestos is given to a number of fibrous minerals that have all been used in the past for either insulation, fire-proofing or reinforcement. Blue asbestos is known to be particularly carcinogenic (when fine particles are taken into the lungs). As a result of wide publicity, the dangers of this material are well known. In terms of existing housing, the main threat is from loosely applied insulation: this should be identified and safely removed by specialists or according to regulations in the government instruction manual. Due to the risk of fire releasing particles into a built-up environment, asbestos in other seemingly inaccessible locations should also be removed when an opportunity arises.

Airborne micro-organisms

Bacteria, mould spores and amoeba are commonly found in indoor air. If they find a suitable place to settle, where the temperature, humidity and nutrients are favourable, they may form colonies, multiply, become airborne and enter the respiratory systems of occupants. Prolonged or heavy contamination can cause allergic reactions in the nasal mucus membrane or the lungs.

Airborne house dust

Apart from the above micro-organisms, house dust contains, amongst other things, hairs and fibres, dandruff, particles of wood, plastic etc., airborne particles from outside, soil particles from shoes and the droppings from dust mites. These will all become airborne if small enough and stirred up by moving furnishings, cleaning and vacuuming. Dust mite droppings, in particular, can pass through a vacuum filter and can cause allergic reactions in some people. You can however buy special vacuum cleaners with filters which do not allow mites etc. through.

Apart from reducing the number of furnishings that carry dust mites, one way of expelling very fine particles to the outside is to use a centralised vacuum system installed on an external wall. Alternatively, there are vacuum cleaners that use water as a filter.

Radon

Radon, a highly dangerous radioactive gas, is mainly dealt with in the **RADI-ATION** chapter. The toxicity of radon comes from alpha radiation, which is particularly dangerous to lung tissue.

Other determinants of air quality

Apart from the natural constituents of air and the commonly found indoor pollutants, there are a number of other important characteristics of air quality:

Relative humidity

Water vapour is an important variable constituent of air. It plays a key role in controlling temperature and weather on the planet. Inside our homes it has an important effect on our health and comfort. The amount of water vapour in the air is expressed in terms of relative humidity, with 100% denoting totally saturated air and 0% totally dry air. The amount of water vapour that can be held in the air increases with temperature, so that very warm air can carry several times as much water as very cold air and still register the same relative humidity level. Very dry air when it is warm causes our mucus membranes to dry out, leading to increased vulnerability to infections, whereas very humid air can cause bronchial problems.

There is a tendency for air to become too dry in winter; there are number of methods that can be used to increase moisture levels. There are various types of humidifiers on the market, ranging from simple trays for water, which you can attach to a radiator, to electric appliances that send a fine mist of water into the air. Alternatively you can use house plants, or change your clothes drying arrangements. Generally for health and comfort it is advisable to keep humidity levels above 20-30%. In order to monitor these levels it is well worth acquiring a small number of hygrometers to place in key locations in your home, such as your kitchen, living room or bedroom.

In certain circumstances humidity levels can become too high. This can be caused by lack of adequate ventilation, with moisture from warm kitchens and bathrooms dispersing to other, cooler parts of the house. In basements, high relative humidity can be caused by damp penetration. Above 70% humidity, mould growth is encouraged, the spores of which can cause allergic reactions. In extreme cases a dehumidifier can be used, but in most cases increasing ventilation rates and attending to the source of the humidity in the first place are the most appropriate solutions. In kitchens and bathrooms, when humidity levels become too high it is worth considering using a humidistat attached to mechanical ventilation, to keep the humidity to preset levels.

Ionisation

Ions are positively and negatively charged molecules. Out of doors, in unpolluted places, the air contains from 1000 to 2000 ions per cubic centimetre, in a ratio of five positive to four negative. This proportion gives a feeling of general well-being. In certain conditions, for instance before thunder storms, the negative ions lose their charge, and a surfeit of positive ions is produced causing unpleasant feelings of tension, irritability, depression and even physical disorders.

Inside the home, negative ions are depleted by electrical fields from TVs and other electrical appliances, static from synthetic fibres (most commonly from carpets) and dry air. Ionisers are cheap units which replace these negatively charged ions; many people find them helpful. Ionisers also have the effect of cleaning the air, as the ions become attached to dust particles which in turn are attracted to surfaces in the surroundings. If you want to use an ioniser and wish to avoid this accumulation of fine dust, it would be worth acquiring one with a filter.

Odour and our sense of smell

Our noses are extremely sensitive organs—our sense of smell is thousands of times more acute than our sense of taste. Air pollution can dull and even damage our sense of smell. An unpleasant smell can be a warning that the air could be unhealthy. In our daily lives, however, we often feel that we have little choice and are persuaded to put up with unpleasant smells. This can cause psychological stress, apart from any possible toxic effects.

We should be able to control the odours in our homes; there should be no need to put up with unnecessary unpleasant smells. We can remove the sources of unpleasant smells and begin to create a more harmonious environment by using natural materials. We can also incorporate smells from sweet–smelling plants.

Ventilation rates

Ventilation is the means by which stale air is exchanged for fresh air from outside. The important measure for this is the number of times that air is exchanged, simply referred to as air changes per hour. The need for ventilation is dependent on a number of factors, which include the number of people using the room, the overall volume, the humidity, the temperature, any sources of combustion, the odour level, the sources of toxicity and finally whether anyone is smoking. The practical aspects of how to approach a ventilation strategy are examined in the **DRAUGHT-PROOFING AND VENTILATION** chapter on p.48.

Use of plants

If you are fond of plants, you have the opportunity to use their many properties to improve air quality. They can be used as humidifiers, to absorb toxins such as formaldehyde, and to add fragrance (see **PLANTS** chapter, p.157).

Priorities for action

1. Reduce the sources of contamination. Eliminate worst offenders, such as open sources of combustion, chemicals and dust.
2. Isolate products of combustion with flues or filters and repair defective heating appliances.
3. When relative humidity levels are too low, use either humidifiers, plants, or even clothes drying, to increase these levels.
4. When humidity levels are too high then consider reducing sources of moisture, increasing natural ventilation, attaching a humidistat to mechanical ventilation or, in extreme cases, using a dehumidifier.
5. Conserve negative ions by avoiding a dry atmosphere and also avoiding synthetic fabrics and carpets. Use an ioniser if you find it helps.
6. Increase natural ventilation and work on finding the right balance between air changes and energy conservation (see **DRAUGHT-PROOFING AND VENTILATION** chapter).
7. Use mechanical ventilation if necessary to exhaust water vapour, fumes or the products of combustion at source.
8. Use plants to help clean the air, moderate humidity levels and introduce fragrance (see **PLANTS** chapter p.157).

Water

There would be no life without water—and James Lovelock's Gaia theories have suggested that there would no longer be water on Earth without life. Water and life are inextricably interdependent. Water is continually being purified and recycled by natural ecological processes, and life is endlessly sustained and regenerated by water. It provides the basic living environment for the majority of life in the form of rivers, lakes and oceans. It is the carrier of nutrients to the plants on which all animals depend and, not least, it is the greatest constituent of our bodies. All human settlements depend on a supply of water, such that it has often been the main determinant of where villages, towns and cities have been situated.

There is now a growing water crisis which increasingly affects us all. Firstly, there is the increasing contamination of groundwater, rivers, lakes and oceans; secondly, there is the over-use of water domestically, industrially and agriculturally, which is leading to altered patterns of flow and the lowering of water tables. The two problems are closely linked. In our homes we can look at:

- Incoming water quality
- Pollution of outgoing water
- Conservation of supplies

Although it is only water quality that has a direct effect on our health, the pollution of outgoing drainage and sewage and the overuse of water are both having an adverse effect on water quality generally. This chapter looks at measures we can take in our homes to help improve the quality of the water we use and the waste water we generate.

Water quality

The quality of the water that enters our homes depends on many factors. There are the natural constituents that have always been a part of natural water sources: the dissolved gases of oxygen and carbon dioxide and the beneficial salts that give water its taste and liveliness. Then there are the additives such as chlorine and aluminium nitrate that are designed to kill bacteria and settle contaminants. After this comes the increasing list of pollutants that are contaminating the sources of mains supply water, whether from ground water, rivers, lakes or reservoirs. Our main concern here are the constituents that are or may be harmful. Here is a list of the more serious offenders.

Nitrates (fertiliser)

Nitrate fertiliser is one of the most worrying of the more recent contaminants. Only half the nitrate fertiliser used in agriculture is taken up by crops; the rest is washed into streams and rivers or seeps into ground water. Since most of our water is taken from these sources there is an almost unavoidable contamination: more than a third of British water supplies contain levels of nitrate higher than EC standards allow. This situation is particularly serious in south-east England. Their effect on our health is indirect since it is inside our bodies that nitrates undergo chemical changes to become nitrites and then nitrous acid, which combines with haemoglobin in the blood and reduces our bodies' access to oxygen. This causes fatigue and in babies it can lead to 'blue baby syndrome'. People with low blood pressure or who are anaemic are particularly susceptible.

Pesticides

Like nitrates, large quantities of pesticides are used in agriculture to kill insects and weeds. A considerable proportion of these runs off into rivers and groundwater, from which water supplies are taken. These chemicals are designed to be toxic in minute quantities and often work by attacking the nervous system. It is still not clear whether or not there is a safe limit for the ingestion of these chemicals.

Lead (plumbing)

Lead is now accepted as a seriously toxic substance; we should reduce our ingestion of it to a minimum. It is one of the toxins that can originate in our housing through the use of lead pipes or the solder that links copper pipes. If your water is soft and acidic it would be advisable to replace all lead and soldered pipework that supplies drinking water taps. You can obtain advice from your local water company on both the acidity of your water and on lead pipe replacement.

Aluminium

Acid rain causes otherwise inert aluminium compounds in rocks and soils to be leached out and into our water supplies. Aluminium sulphate is also added to our water supplies to settle out solid impurities. Aluminium is thought to be implicated in Alzheimer's disease. Although water companies are managing to reduce levels of aluminium to reasonable concentrations in our drinking water, its effect on our body is cumulative and we should do what we can to cut down the total amount we ingest. Limiting our use of aluminium cooking utensils is perhaps the most important way of achieving this.

Solvents

There are a growing number of solvents entering the water supply (see list on p.125); it is not yet clear how seriously they affect our health, nor how effective various treatment processes are. Watch for the results of research if you are aware of this being a problem in your area.

Microbial contamination

This type of contamination of mains water by microbes such as bacteria or viruses is rare, due to the use of oxidation techniques and chlorination employed by water companies.

Chemicals added during the treatment of water:

Chlorine, which gives modern tap water its distinctive taste, is used by water companies to kill bacteria. However chlorine also combines with natural acids in water from peaty soil and decomposing vegetable matter to form trihalomethanes (THMs) of which chloroform, a toxic compound, is one. Chloroform can be released into the air in a hot shower spray. The amount of THMs in water supplies varies considerably, and water companies are working to solve this particular problem.

Fluoride

Fluoride is added to our water supplies by some water authorities because of the effect it has on hardening the enamel on teeth and thereby reducing tooth decay. However, fluoride is also toxic and the decision to add it to water supplies remains a controversial and emotive topic. Some scientists have linked the use of fluoride to cancer and genetic damage and the balance of opinion internationally seems to be moving against its use in public water supplies.

Water analysis

Proper and accurate water analysis is a difficult and exacting task and it is thus very expensive to have samples of your own tap water analysed. However, your local water company should be able to provide you with a breakdown of at least the main constituents in your water as it leaves their plant.

Drinking water

Our bodies require one to two litres of water a day depending on activity, climate and humidity, etc. Add to this the water used for cooking and washing vegetables and fruit and we arrive at an average figure of around 5 litres per person per day that is related to eating and drinking. It is obvious that this water should be of the highest standard of any that we use but, unfortunately, a standard of quality suitable for drinking cannot be guaranteed by most water companies. There are three possible responses to this situation:

- To lobby for improvements in the quality of drinking water delivered from your local water company
- To use bottled water and
- To use a domestic filtering system

The first solution requires some understanding of how improvements can be made to public supplies and is beyond the scope of this book. We can also reduce the amount of pollution that we add to the waste water system (for example by not putting paint and cleaning solvents down the sink). By buying organic produce, even if at greater cost, we can help to reduce the amount of pesticides and synthetic fertilisers used in agriculture.

The second solution, to use bottled water, has huge ecological disadvantages as its supply involves huge transport costs and the use of non-recyclable bottles. There are no standards for bottled water, so there is no guarantee that it will not contain contaminants. Also, plastic bottles themselves often contaminate the 'spring water' they contain.

Water filters

As things stand, there is little alternative but to use a filtering system for our drinking water, if we want to improve its quality. The simplest method is to use a filter jug which you can buy in the high street. It is important to choose one that is recommended for the particular pollutants that are likely to be present in your supply. However if you want greater ease of use, you need to install a plumbed-in system.

There are three basic types:

1. Activated carbon

This is the least expensive and most common of the three types of filter. It simply passes the water through carbon granules or a compacted synthesised cylinder of compressed carbon powder. This type of filter removes chlorine, pesticides and organic chemicals, but not

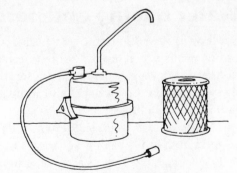

Carbon-based water filter.

nitrates. Some (but not all) heavy metals are removed. The filter requires changing relatively frequently before it becomes clogged with pollutants. If this is not done, bacteria can be attracted to it and pollutants can be released back into the water. Alternative filters that include granules which remove nitrates are becoming increasingly common.

2. Reverse osmosis

The principle of this filter is a semi-permiable membrane made of cellulose through which the water is forced under pressure. This filters out most pollutants with large molecules, only letting smaller molecules through. Although it filters out all particles, bacteria, aluminium, heavy metals, minerals, salt and nitrates, it does let many volatile organics through. It also has the disadvantage that it filters out the desirable minerals of calcium and magnesium.

3. Distillation

This system is sold as one that parallels the water cycle in nature. However it is extremely energy-consuming to run as all the water needs to be turned to steam before being condensed. Unless the heat can be effectively used elsewhere, or solar power can be used, from an ecological point of view it should be ruled out. Distilled water is flat to taste; and some volatile organic chemicals condense with the water.

The judgement to be made about which system to choose will almost certainly depend on the choice of systems you can track down. The detailed specifications of any system will need to be matched with the particular pollutants in your local water supply. Most of the systems available in Britain are the activated carbon type. Some addresses of suppliers are given at the back of the book. Since there is no British Standard for filters, ask for laboratory test reports or check consumer magazines to make comparisons.

Water quality and personal hygiene

Swallowing water is not the only health risk when it comes to the the quality of our mains water supply. Some research has found less chemical absorption from drinking contaminated water than from using it to wash or take a shower. In these cases chemicals can be absorbed in two ways: through the skin and via the lungs.

Through the skin: chemicals can be absorbed while in the swimming pool, taking a hot bath, washing up or washing clothes by hand.

Via the lungs: researchers, experimenting with showers and common water pollutants, found that certain toxic chemicals evaporated into the surrounding air. Increasing concentrations of the chemicals also gradually built up in the shower and spread through the home. The amount of the chemical that vaporises increases with more powerful and hotter showers.

Some researchers believe that 50-60% of soluble contaminants that we absorb enter the body through the skin or lungs in the ways described. Should we then filter the water for baths and showers? The answer will depend on how serious the volatile pollutants are in your area as well as how often, long or hot, you take your baths or showers.

Reducing our own water pollution

The water that enters the house relatively clean leaves it in various states of contamination from organic waste such as urine, faeces, paper and food waste, and all the chemicals we use: washing powder, bleaches, washing-up liquid, dishwasher powder, toilet cleaners and fresheners, DIY paint solvents and cleaning liquids, etc.

The main consideration is the biodegradability of particular chemicals. There are now 'green' products of most types of washing powder or cleaning liquid and there are green consumer magazines which help to determine the differences between them. In many cases we can reduce our usage or dispense with these chemicals altogether.

Composting toilets

Another way of increasing your own self-sufficiency and reducing outgoing waste water, especially if you are not attached to a mains sewer, is to use a composting toilet. These are gradually gaining favour in many parts of the western world, particularly in the United States and Scandinavia. There are a variety of different systems and they all have one thing in common: no water is used.

- The Clivus Multrum (Sweden) model depends on having space directly below the toilet and ideally the kitchen so that both faeces and organic waste from the kitchen drops into a composting chamber. A small fan and associated ducts promotes air flow to ensure odourless aerobic decomposition.
- The Sun-Mar System (USA) manages to reduce the space required by the Multrum so that it is possible to have the whole system in one unit. A small step allows you to sit on the composting unit itself. A 'Bio-Drum' ensures effective aeration and sterilisation and speeds up the process by rotation. This is activated by turning a small handle. The

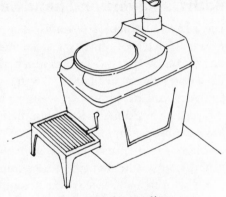

Composting toilet.

great advantage of this system is that it can be put almost anywhere and requires only a vent pipe.

- The Lectrolav (UK) avoids composting within the unit by dehydrating both solids and liquids electronically to a fraction of their original volume. These dehydrated solids are then removed for composting elsewhere.

We can also treat our own waste water organically with plants if we have the necessary external space and are independent of a local sewage system (see PLANTS chapter p.157).

Water conservation

Apart from reducing our waste water contamination, we can also reduce the amount of water we use. While the overall amount of water used by industry is decreasing, the amount used by households and agriculture is increasing steadily. On average each person uses in all about 135 litres (30 gallons) of water a day. Below is a table giving approximate average percentages of water used in different activities:

	%
Flushing Toilet	30
Personal Hygiene	25
Laundry	12
Washing up	8
Drinking and Cooking	5
Watering Garden	5
Washing Car	2
Losses	13
Total	100

Baths, showers and handwashing

If we have a choice between using a bath or a shower, a short shower will generally save a lot more water. Saving hot water will save energy as well. A standard shower can deliver as much as 14 litres of water per minute. If you use a water-saving head instead, water consumption can be halved. The same approach can be applied to washing hands—various types of aerators and sprayers are available. In addition, the less water that is used for washing, the less our absorption of chemicals.

There are of course other ways we can vary our washing habits and save water, energy and chemical absorption:

- Swim in a sea, lake, river, stream or pool
- Take a cold shower

- Wash with a bowl of water and a jug.

There are countless ways that people have used to wash themselves across the world and down the ages. Western society has sometimes become obsessed with body cleanliness beyond reason. This obsession could now be turned to the cleanliness of our polluted environment.

Washing machines

Washing machines use between 70 litres and 115 litres of water per load. This represents about twice the amount you would use if you washed by hand. There is plenty of scope for reducing the amount of water that washing machines use and for improving the energy-saving and water-saving programmes. There are big differences in water consumption for different models of washing machine and this should be a major consideration when buying a new one. One model, in a *Which?* magazine test, used around 160 litres on a hot cotton wash.

Dishwashers and washing up

The same applies to dishwashers as for washing machines. Most dishwashers are very wasteful and use far more water than necessary. Compared to washing machines, however, there is even greater potential to be efficient with both energy and water, if a dishwasher has been carefully designed.

Water closet

The toilet is the most wasteful of all household conveniences in terms of water: this 'convenience' takes about 20 litres of tap water, mixes it with excreta and paper and flushes it into the sewer. For some houses a composting toilet is a possibility (see p.135) but for most it will not be a high priority to replace an existing functioning system. We can, however, conserve water by either reducing the amount of water flushed to a minimum, or replacing the toilet itself with one of a low flush design. We may also feel we needn't flush every time. Another step to take if water conservation is a main concern is to use 'greywater' from showers, baths or washing to flush our toilet (see WASTE WATER RECYCLING overleaf); an ingenious design (see illustration), allows us at least to wash our hands in the water before it is used for flushing.

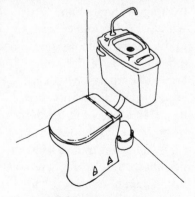

Basin-topped cistern.

Waste water recycling

There are three different categories of water that are drained from our houses: Rainwater, sewage and greywater from washing clothes, dishes and bodies. There are many arrangements for recycling greywater and/or rainwater. They all have these features in common:

- Waste water pipework collecting greywater from baths, shower, basins, washing machines and any other source of greywater.
- A simple filter to remove any larger solid particles.
- A storage tank to store the waste water.
- A distribution system to take it to where it is needed—toilets or a garden tap for watering or washing the car. Either a pump or a gravity system can be used for this distribution system. Drip irrigation can also be incorporated for watering the garden.
- Rainwater can either be an integral part of this system or treated separately.

The practicality of using such a system in your house is very much dependent on your existing arrangements. There will also be differences in terms of the need for such systems in different parts of the country. For those that have experienced restrictions on water usage and have big gardens, there are very obvious advantages. The illustration alongside shows the use of greywater from upper stories to feed toilets on the ground floor.

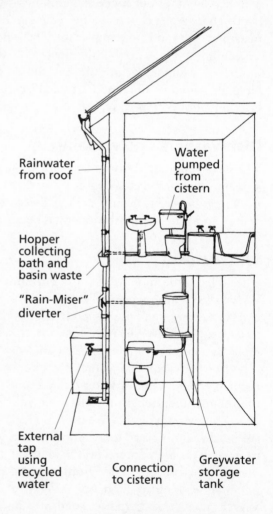

Rainwater from roof

Hopper collecting bath and basin waste

"Rain-Miser" diverter

External tap using recycled water

Water pumped from cistern

Connection to cistern

Greywater storage tank

Rainwater collection and greywater recycling.

Priorities for action

1. Obtain a water analysis from your local water company and determine what measures you may need to take, either in filtering or lead pipe removal.
2. Consider fitting a plumbed-in filter system (depending on the quality of your water). Match the system to the pollutants you wish to filter out.
3. Reduce chemical pollution carried from your home in waste water by using biodegradable versions.
4. Reduce the amount of water used in showers, baths and handwashing (see **WATER CONSERVATION** on p.136)
5. If buying a washing machine or dishwasher, choose one with a low water consumption programme.
6. Consider exchanging your toilet and/or cistern for a more efficient design.
7. Consider a system of storing greywater and/or rainwater so that it can be used for toilets, for watering the garden or washing your car.

Radiation

It is sometimes difficult to appreciate just how many different types of electromagnetic radiation impinge on us and our environment. This is largely because, with the exceptions of light and heat, we cannot see or feel them. The electromagnetic spectrum comprises a large range of radiation with very different wavelengths and properties, but many aspects of this radiation are little understood, and, still less, what effect they have on our general health and well-being.

This chapter looks at the most important types of radiation that are known to affect us in our homes, and in the renovation of our houses. First, ionising radiation and its main source (radon) are looked at, as are the ways we can keep it from entering our homes. Secondly, non-ionising radiation in all its various forms—ultraviolet and visible light, heat, microwaves and radio waves—are analysed in terms of their known effects on our health.

Table of the Electromagnetic Spectrum

Approx. wavelength in metres	Type of Radiation
10^{-23}	Cosmic radiation
10^{-20}	Gamma rays
10^{-19-18}	X-rays
10^{-17}	Ionising ultraviolet (UV)
10^{-15-16}	Non-ionising UV
10^{-14}	Infra-red (heat)
10^{-10}	Microwaves
10^{-5-9}	Radio waves
10	Schumann field (electromagnetic pulse)

The table above gives some idea of the different types of radiation within the electromagnetic spectrum. All electromagnetic radiation consists of energy waves travelling at the speed of light. It is very difficult for us to visualise these waves as their properties are mostly outside our normal experience. Our best model of radiation is visible light, as we can follow its path; this gives us a little insight into how these waves behave. When this radiant energy strikes matter it can, if strong enough, permanently alter its state.

The most important distinction for health within the electromagnetic spectrum is that between ionising and non-ionising radiation. In the case of ionising radiation some of these waves—the alpha and beta waves—are in the form of charged particles which can collide with atoms and molecules, knock out electrons and form ions. The other ionising rays consist of gamma rays, x-rays, and part of the ultraviolet spectrum.

Non-ionizing radiation, on the other hand, consists also of energy waves which act on matter by transferring energy, usually in some form of vibrations or heat, but non-ionising radiation does not have the ability to change the structure of atoms. Its wavelengths are larger and act on a supra-atomic scale. Mild non-ionising radiation simply creates resonances which can be more subtle and unpredictable in their effect.

Ionising radiation

Ionising radiation is the most dangerous type of radiation for living organisms although, paradoxically, without its presence it is unlikely that life would have come into being and evolved in the first place. When cells are affected by ionising radiation, various degrees of damage can be caused. With a strong dose, cells can die or fail to reproduce. Ionising radiation is, after all, used medically to kill cancerous cells. With a weaker dose, genetic material in the cell can be altered or its biochemical mechanisms affected. In most cases a healthy body will identify sick or malfunctioning cells and destroy them, but sometimes damaged cells may survive and go on to reproduce abnormally. This can result in altered cell behaviour, cancerous growth or, if the damage occurs in genes of the reproductive cells, genetic mutations.

There are many sources of ionising radiation that we live with:

From within our bodies:

- Radioactive elements that have always occurred naturally in our bodies such as radioactive potassium
- Radioactive elements in our bodies that are a result of the products of (man-made) nuclear fission such as strontium 90

From our environment:

- Radiation from natural sources on the planet, such as uranium and radon, and from building materials containing such radioactive elements
- Radiation from the fallout from man-made nuclear explosions and power station accidents
- Radiation from man-made sources within our immediate environment such as fluorescent dials, TVs, some smoke alarms and X rays (these can be categorised as personal, domestic, industrial or medical sources)

From the cosmos:

- Radiation from the cosmos – the sun and cosmic radiation. Most of this is shielded from us by molecules in the atmosphere, such as ozone. Small amounts do penetrate, however, such as ionising ultraviolet radiation.

In our homes our main concerns are:
- Radioactive elements in the ground and building materials—in particular the gases radon and thoron that are given off by these elements
- Ionising radiation from domestic appliances— some smoke alarms and cathode ray tubes (televisions and computers)
- Ionising radiation from the sun (see ULTRAVIOLET on p.145)

Radon

Radon is a natural radioactive gas which comes from the decay of uranium via radium. As there are traces of uranium and radium in all soils and rocks, radon is continually being emitted from the ground. It has a radioactive half-life of 3.8 days. When it is emitted into the open air it is rapidly diluted, but if it enters a basement or underfloor cavity its concentration can build up if it is not ventilated away. People exposed to high radon concentrations for long periods have an increased risk of lung cancer. It has recently been discovered that it is the radon molecules that lodge in the lungs that cause the damage by emitting very short range but deadly alpha rays. It has now been estimated that in Britain as many as 2,500 deaths from lung cancer a year are a direct result of radon exposure. Thoron is another radioactive gas that people are exposed to but its half-life is only one minute and so it does not diffuse far from its point of origin.

Radon from the ground

There are four factors which affect the concentration which can build up within a building:
- The concentration of radium in the ground. This varies enormously, from sedimentary rocks to ground containing uranium ore. The highest concentrations in the UK are found in south-west England, where the source is granite.
- The permeability of the ground. Sand and gravel provide the most permeable conditions whereas heavy clay and rock provide the opposite.
- Penetrability of the building. Cracks, construction joints and gaps around service pipes all provide pathways for the gas.
- Reduced air pressure in buildings. The air in buildings is frequently at a slightly lower pressure than outside owing to the stack effect and extraction fans. This has the effect of drawing the gas into the building (see p.49 of DRAUGHT-PROOFING AND VENTILATING).

Radon will only be a major cause for concern if you are living in a high-risk area (mainly parts of Devon and Cornwall) and you will probably have been informed if you are. If you think you might have a problem, however, contact Radon Survey, NRPB, Chilton, Oxon OX11 ORG, who currently provide a free measurement service. Grants and advice are also available to make the necessary alterations. The only way to tell reliably if you have a problem in a particular part of the house is to place a passive radon detector there over the period of a few months.

What to do if you find you have a radon build-up

Since radon gas is only likely to accumulate in unventilated basements and floorspaces that have a direct pathway to the ground, it is the ventilation and sealing of potential radon routes in these areas that you should be concerned about. In most houses with ventilated underfloor spaces, it is a relatively simple process to increase the ventilation to this space and seal the floor between. If you have a basement or cellar that you visit infrequently, you can treat this as though it were part of the ventilated space under the house: ensure it is well ventilated and that the door to the cellar or basement is properly draughtproofed. If on the other hand you have solid floors, or a basement that is used for sleeping or day to day living, there is a probability of radon infiltration. There are three approaches to this problem:

- Install a vapour-proof barrier in the floor and any walls below ground level.
- Ensure adequate ventilation of the living areas.
- Reduce the pressure of the radon gas in the ground where it emanates from, immediately below the floor concerned. The Building Research Establishment has recently produced a booklet (see **REFERENCES**) describing the installation of radon sumps as a way of achieving this. Air is sucked out of the sump by a fan thus reducing the chance of radon being pressured through small cracks. The illustration opposite shows possible arrangements for doing this.

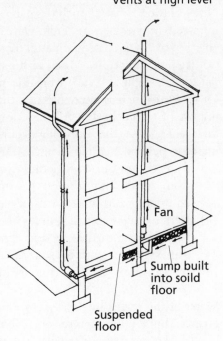

Vents at high level

Fan

Sump built into soild floor

Suspended floor

Dealing with radon.

Radon from building materials

Virtually all building materials contain uranium and so emit minute amounts of radon. Building materials also emit gamma rays. In most cases the amounts are so small as to be negligible, but it has been found that certain building materials such as crushed granite bricks and blocks do emit significant amounts of radon. If you are concerned that your home is made from these materials you can obtain a free measurement survey and advice from The Radon Survey (see above). The same measures should be taken as with basement areas, i.e. ensure a vapour-proof barrier is used on the inside of the walls containing the radioactivity. Certain paints can be used to provide this barrier.

Radioactivity from household appliances

Some smoke detectors have a radioactive source: danger only occurs at close range over long periods of time. By ensuring that smoke detectors are placed on ceilings well away from beds or sitting areas this danger can be almost completely eliminated. With TVs and computer screens the ionising radiation is very small, but it is advisable to limit prolonged and close exposure to this type of equipment.

Non-ionising radiation

Non-ionising radiation includes part of the ultraviolet spectrum, visible light, infra-red, microwaves and radio waves. All these forms of radiation consist of energy waves which act on matter by transferring energy in the form of electromagnetic vibrations. Much of the natural non-ionising radiation is essential for life: light and heat from the sun being the most obvious examples. Other forms such as ultraviolet and the Earth's electromagnetic pulse are also important to our health. All these forms of radiation are beneficial to us in the average doses that normally occur, but become harmful if their intensity increases beyond a certain point. Sources of non-ionising radiation are:
- from within our own bodies (mainly heat)
- from inside our homes: lighting, heating, microwave ovens, TVs, computers and most forms of electrical equipment when in operation
- from natural external sources: the sun, the cosmos, the earth's magnetic field and the earth's electromagnetic pulse
- from synthetic external sources: radio and TV broadcasting, microwave transmissions, electricity generation and distribution systems and radar

Many of the forms listed above are radio waves and there is no doubt that given a high enough intensity even these wavelengths can be harmful; however in most cases the level is extremely low. It is difficult to be able to predict from our present knowledge that a particular radiation field will produce any particular effect on our bodies. However, since so many of the processes within our bodies are dependent upon electromagnetic interactions between molecules, it would indeed be surprising if these fields did not have some effect on our central nervous system. Research in this area is developing fast and more precise information should be available in the near future.

Ultraviolet radiation (UV)

Ultraviolet covers a large range of wavelengths—ten times that of the range of visible light. Just as visible light has different properties for different wavelengths so has ultraviolet, with the shorter wavelengths being more destructive and having ionising properties. At present we seem to be much more concerned with this ultraviolet component of sunlight which causes skin cancer than with the longer wavelengths that are actually beneficial to our health. Contrary to what we may think, regular doses of sunlight generally appear to help prevent cancer (see *Day Light Robbery* by Damien Downing in the REFERENCES at the end of the book). What is it then that causes skin cancer? It is simply sunburn, which is an overdose of solar radiation containing destructive ionising radiation. This overexposure to ultraviolet light is the same as with many wave energies: a mild dose is positive, a medium one is stimulating and a strong one harmful. Before World War II it was generally accepted that being out in the sun was healthy—hospitals had verandas and balconies on which patients could receive the health-giving rays. Besides producing vitamin D, ultraviolet also cleanses the blood by killing bacteria in it as it passes through our surface blood vessels. It also reduces the risk of heart disease and some cancers.

The depletion of the ozone layer

The effect of the destruction of the layer of ozone molecules in the upper atmosphere, primarily as a result of the release of chlorofluorocarbons (CFCs) into the upper atmosphere, is that less and less of the damaging ultraviolet rays are filtered out. There is uncertainty about quite how this will effect the biosphere as a whole, or human beings in particular. However we in the northern latitudes, hidden away in our homes for most of the time, on average need to *increase* our overall exposure to UV radiation. Our problem is that we tend to go for a whole year's supply on one sunny bank holiday weekend or during a week in Spain. This leads to very intense exposure for vulnerable skin, the cells of which can be seriously damaged.

The filtering effect of glass

Glass filters out UV light, which is useful in that it prevents the degradation of materials inside the home, but it is not so useful for our health. If we could build conservatories which were transparent to the healthy wavelengths of ultraviolet, we could sit in the sun even on cold days and soak up our weekly dose of ultraviolet light. Such glass used to be available before the war but is no longer obtainable. However most plastics do allow ultraviolet to pass through, so an alternative could be to glaze over part of our conservatory or even balcony with a suitable type of plastic pane. Certain plastics are being increasingly used for glazing.

Full-spectrum lighting

Another way of exposing ourselves to ultraviolet light while indoors is the use of full spectrum lighting. This is lighting which produces a mix of wavelengths similar to that of the sun, including ultraviolet radiation. If we were to include this type of lighting in rooms such as bathrooms or bedrooms, where we remove our clothes, we could increase our dose of ultraviolet light, particularly in the winter when we most need it. At present full spectrum fittings are available in tubes and housings similar to the old-style fluorescent ones. Since all fluorescent lighting is based initially on the production of ultraviolet it should be possible to produce fittings in any of the compact versions now available (see LIGHTING on p.86)

Visible light

Most energy from the sun is radiated in the form of light. Plants developed photosynthesis in order for them to make use of this most abundant source of energy and our eyes developed sensitivity to it for the same reason. Sunlight thus plays a very important part in our well-being.

Seasonal affective disorder (SAD)

Sleepiness is partly caused by the hormone melatonin, which is secreted by the pineal gland in the brain during darkness or dim light. However if we are in the dark or in dimly lit interiors for long periods, melatonin is over-produced and can cause depression or seasonal affective disorder (SAD). If someone is suffering from this syndrome—and it may be something we all suffer from to a greater or lesser extent in winter—therapy for this condition is bright light. Bright light in general has the effect of making us feel more positive and awake and for this reason we need to ensure there is plenty of natural light in our home (see LIGHTING chapter). This provides yet another reason for having a conservatory or sunroom which we can use when we feel the need.

Glare

There are various forms of poor lighting that can cause problems for the eyes:

they can increase stress, and cause headaches and eye strain. The most common problem is that of glare. The eye adapts to the average light intensity of what it is looking at, but if there is a very bright area in the field of vision then discomfort is experienced. There are three types of glare that are commonly identified:

- *disability glare* is caused by direct, intense light affecting the ability of the eye(s) to view an object.
- *discomfort glare* arises from long-term glare from windows and light fittings that are in the field of view.
- *reflected glare* is common where the object is shiny and either a window or a light is reflected from its surface.

Glare can usually be avoided by lighting the task more brightly (2 or 3 times) than the immediate surroundings, with the further background even less bright. Extremes of brightness should be avoided.

Colour

The effects of different colours on us in terms of health are not fully understood. However since colour is such an important part of light, it is worth looking at some of the effects that have been observed. Certain themes run through and are common to many studies.

Effects of different colours

Blue	Calming, cooling and relaxing
Green	Harmonious and healing
Yellow	Cheering & stimulating to the intellect
Orange	Stimulating to the intellect and emotions
Red	Emotionally stimulating
Pink	Considered a colour of universal healing
White	Neutral

Some of the results of research suggest that colours work on our subconscious. However, colours correspond to different bands of wavelength within the light spectrum and also correspond to different neural stimuli in the eye. There is some evidence to suggest that different parts of the brain are stimulated by the different wavelengths of different colours. As with so much psychological research, separating the psychological from the physical is difficult. At present the main findings remain empirical and, as might be expected, there are large cultural and personal variations. For some people colour is very important and has a clear influence on their state of mind. If you are one of these people, use the colours in your home that feel right for you.

Infra-red

Infra-red radiation is emitted by every single object in proportion to its temperature. Heat is essential to life and radiated heat is one of the major factors that makes us feel warm or cold—this is further dealt with in SPACE HEATING on p.70. We are well protected by our senses when it comes to heat as our skin is very sensitive to temperature.

Infra-red wavelength controls are increasingly being used in domestic environments to turn lights on and off and provide security systems. Like visible light, infra-red is an essential and integral part of our environment. At low intensities it is completely benign.

Microwaves

Our potential exposure to microwaves is mostly through microwave cooking. (see APPLIANCES on p.94). If you do have a microwave oven it is important to understand the effects of microwave radiation if it 'leaks'. There are two ways that microwaves can affect the body:

- By direct heating of the body and
- By non-thermal effects

The first is straightforward. Direct heating takes place, and if too intense will overheat or burn the exposed parts of the body. The eyes are particularly vulnerable. The non-thermal effects are less clear but there seems to be strong evidence that they can be harmful; we should guard ourselves against unnecessary exposure by keeping door seals to ovens in good condition.

Radio waves

Besides those produced by radio transmitters, radio waves of different types are produced by most electrical installations. In fact all electric current flowing through cables or wires creates radiation: the larger the current the stronger it is. Power transmission lines carry very large voltages and emit correspondingly high amounts of radiation. There is now some evidence that there is a link between proximity to these power lines and the incidence of leukaemia and cancer. Children are the most affected, and the danger seems to be at its greatest when people sleep close to these cables. How much we are affected by the electric cables in our homes is more difficult to determine; however some scientists advise reducing or eliminating electrical equipment near your bed, or shielding electric circuits. One reason radiation from electronic equipment and circuits can be a health risk is that much of it is in the form of alternating current.

Alternating and direct current

All domestic supplies that come from central grids are alternating current (AC). Unlike direct current (DC), where the current is continuous in one direction, AC reverses its flow continually in an alternating cycle. This frequency is 50 cycles per second (50 hz) in the UK (60 hz in the USA). Alternating current is the preferred form of electricity supply for numerous reasons: it can be transmitted over long distances at high voltages with little loss of power; and it can be converted easily from one voltage to another, or to DC, as required by the end user. However this alternating frequency gives rise to radiation, which coincides with frequencies that are used by the brain. It is still not understood what effect this has, but there is evidence to suggest that some people's mental abilities are impaired. This is a rapidly growing field of research and may prove another reason for reducing our excessive dependence on high voltage electrical power.

Schumann waves

Schumann waves are a natural from of radiation emanating from the Earth's electromagnetic field, which pulses at the rate of 7.83 beats a second. These waves are thought to be essential to our well-being, partly because the human body's bio-electrical system also pulses at about the same rate and also because when astronauts become separated from this pulse they experience a loss of orientation. This pulse is now provided synthetically on all spacecraft to ensure the health of the crews. What is not so clear is how the earth's pulse may be masked or deflected by buildings and metal components in buildings. It is clear that some people find that they cannot sleep well in particular locations and orientations, although this may be for reasons other than a lack of Schumann waves.

Priorities for Action

1. If you live in a high risk radon area and have not yet taken precautions, contact the Radon Survey for a free measurement and undertake the measures outlined on p.143 if necessary.
2. Consider building a conservatory or covered balcony with glass or plastic that allows ultraviolet through to provide a healthier environment for you and your plants.
3. Consider installing full spectrum lighting in your bathroom or bedroom.
4. If you have a choice, arrange your rooms to follow the sun to make the best use of bright sunlight, especially in winter to reduce seasonal affective disorder.
5. If you have an old microwave check whether its door is sealed properly. A damp cloth held in front of the appliance will warm up where leakage is occurring.
6. If you have high voltage cables or transformer stations nearby, you can use a portable radio to 'listen' to the interference produced. If you can pick up any of this interference in your home, sleep in one of the rooms which are least affected.

Sound

Sound is a neutral term which encompasses both noise which we find a disturbance and an intrusion, and pleasurable sounds which we generally welcome. It is noise, and how to control it, that we are most concerned with in this chapter. How we experience noise is mostly subjective: it often depends whether we are making the noise or on the receiving end of it.

How our bodies react to sound

Sound impinges on our bodies as a series of pressure waves or vibrations which vary in intensity and frequency. The intensity (or pressure) of the waves determines the loudness of the sound and the frequency determines the pitch (how high or low it sounds). When these pressure vibrations reach our body, they set up vibrations in all the cells immediately adjacent—if the sound is loud enough the vibrations will travel to all parts of our body. There is evidence that each cell in the body is affected by the vibrations of sound. The ear is of course the organ that makes sense of these waves—it is able to pick up an incredible range of frequencies and loudness. These are transformed into nerve signals for the brain to decode and interpret.

The range of response of the ear is truly amazing. It is possible for the human ear to pick up sounds that are one billionth the strength of the loudest sounds that can be heard without causing damage. Because of this vast range, a logarithmic scale is used to measure sound. This decibel scale starts with 0 decibels for the threshold of hearing, assigns 60 decibels for normal conversation and 120 decibels for the threshold of pain. A chart below gives some typical noises with their approximate decibel rating:

Sources of Noise	Perceived decibels
Jet engine at 30m	130–140
Pain threshold – loud thunder	120
Discotheque speakers at 2m	110
Motor cycle at 2m	100
Heavy truck	90
Danger level if noise continuous	80
Vacuum cleaner	70
Normal conversation	60
Inside average home	50
Refrigerator	40
Whispering	30
Swallowing	20
Rustling leaves	10
Hearing threshold	0

Levels of sound fall into four categories:
- Background sound (ambient)
- Normal (healthy) sound levels
- Danger levels (for continuous exposure)
- Painful or damaging levels

Like other forms of sensory awareness, there is a level of stimulation which is both healthy and necessary for well-being. Below this, we suffer from sensory deprivation. However with sound there is almost always background noise no matter where you are. It is only in deep space or in specially constructed studios that it is possible to experience no environmental sound at all (it is impossible to eliminate the sounds from within our own bodies).

Above this basic background level comes the normal sound range with which we usually deal. It is within this range that we find the levels of speech and the sounds we find harmonious, such as most music and most of the sounds of nature. Within this range are also the sounds that we find socially unacceptable or embarrassing, including those we hear through the walls from our neighbours as well as bodily noises that we know people do not wish to hear. Mostly it depends on how our brains label these sounds. The more they are labelled as annoying, the more stress they are likely to bring. Sensory fatigue is another form of stress which occurs with a constant unwanted noise, especially if repetitive. Think of the simple drip of a tap during the night.

Danger levels of noise are those at which our ears can be damaged with prolonged subjection, particularly if at one single frequency. Such levels of noise cause permanent damage to our ears but are rarely encountered in domestic circumstances (except when listening to music too loudly with headphones).

Most solutions to noise problems that occur in our homes will fall into one of the following four categories:
- Avoiding making noise that causes annoyance to others – which usually means controlling a noise at source.
- Reducing noise that travels from one part of the house to another, to give greater privacy and reduce annoyance.
- Reducing noise from neighbours that is transmitted through a party wall.
- Reducing noise from external sources such as road or air traffic.

Control of noise in your home

In order to be able to begin to control the noise in your own home it is necessary to understand how noise is created and transmitted:

Airborne sound

Noise is most easily transmitted if there is an uninterrupted pathway, for instance from outside to the inside through open windows, gaps around doors or even through small cracks. To reduce airborne sound it is simply necessary to seal gaps in the element that you wish to act as the sound barrier.

Impact sound

This is the type of sound that is produced by the impact of shoes or furniture on floors, or by a masonry drill drilling a wall. Heavy objects and hard surfaces produce the most impact sound, so by using soft shoes and soft floor coverings this type of sound can be reduced.

Structure-borne sound

Impact sound is carried through the walls and floors of your home by the structure itself, which carries the vibrations and transmits them to various parts of the house. Structure-borne sound can be reduced by introducing discontinuities, such as the introduction of a resilient layer. An example of this is the floating floor, where flooring is laid on resilient pads or matting, such as rubber underlay, in order to isolate the floor surface from the structure (see illustration on p.155). Noise from the use of stairs is a common structure-borne sound.

Panel sound

This is a particular type of transmission that occurs when a whole partition or pane of glass is made to vibrate like a drum. It is most likely to occur with lightweight partitions. The transmission is also greatly increased if the noise has a frequency similar to the natural resonance (or vibrating frequency) of the panel. This type of transmission can be reduced by increasing the weight of the partition in such a way as to damp the vibrations. Panes of glass can also act in this way, and in this case secondary glazing will reduce noise transmission considerably (see p.156).

Reverberation

Reverberation occurs when sound is reflected backwards and forwards between hard-surfaced walls: a completely empty room has a particular hollow echo. Sometimes we like a room to have some reverberation, such as when playing music—or even when singing in the bath. However it is normally more restful, especially with children, to have rooms that act to deaden the sound. This is usually easily achieved with soft furnishings such as carpets, curtains and furniture.

Reducing noise at source

Theoretically, you have control over noises that emanate from within your own home. Examples of noises that people find annoying are those from garden or kitchen machinery, amplified music, noisy stairs, noisy plumbing, banging doors or DIY construction. Some examples will help you to see that there is almost always something you can do to reduce any kind of noise, if you are prepared to do something about it.

A *washing machine* produces:
- noise directly from the machine itself
- impact sound from vibrations on the floor; and
- structure-borne sound from these vibrations.

The noise of the appliance itself can often be reduced by damping the metal panels making up the carcass. One way of achieving this is to glue rubber or foam to the inside of the panels, provided this does not interfere with the running of the machine. Alternatively the machine could be partly or wholly enclosed to help isolate the sound.

The noise from the vibrations on the floor are helped by placing the feet on thick rubber bases to absorb the vibrations. If there is still a problem with structure-borne sound, the whole machine could be placed on an insulated raft to further isolate the vibrations from the main structure of the building.

Noisy stairs: the sound can be reduced by using a carpet or other covering with a resilient foam rubber underlay.

Noisy plumbing: this problem can often be solved easily if the cause is correctly identified and action taken. If you cannot do this yourself, find a plumber who is interested in dealing with such problems. It is often the result of too sharp a bend in a small pipe. Restraining vibrating pipes effectively can also help to reduce the problem.

Banging doors can be prevented by laying draught-proofing strips in the rebate.

We can also exercise control over many aspects of the noise we produce, such as the volume of the music we listen to, and the loudness of our voices.

Reducing the transmission of noise

The commonest examples of noise problems within a home are:
- Noise being transmitted through the floor to the room below
- Noise being transmitted through a lightweight partition to an adjacent room.

The most effective solution to the first problem is to lay a new floor, such as chipboard or ply, on top of a resilient underlay, ensuring that the edges of the new floor do not contact the walls or skirting. A soft floor covering can be further laid on this new surface. This means of soundproofing is illustrated below.

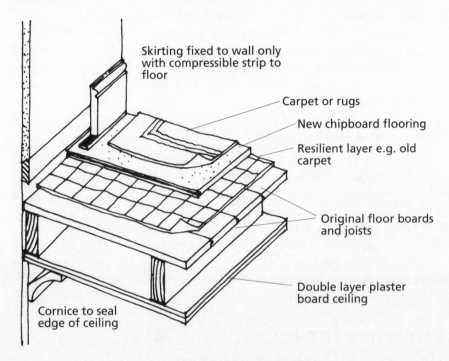

Skirting fixed to wall only with compressible strip to floor

Carpet or rugs

New chipboard flooring

Resilient layer e.g. old carpet

Original floor boards and joists

Double layer plaster board ceiling

Cornice to seal edge of ceiling

Sound deadening to floors.

The problem of reducing noise which is transmitted through a light weight partition can be solved by increasing the effective weight of the partition. This can be done by adding a layer of plasterboard to the partition, fixed with an acoustic sealant to help damp the sound. Alternatively, fitted shelving or cupboards can be used to add weight.

Reducing noise through a party wall

This is the most difficult noise to deal with: it has already carried through a heavy party wall, and the measures applied to a lightweight partition will have less effect. However the following measures should help:
- Check that there are no cracks in the wall, as these will tend to weaken the acoustic integrity of the wall.
- Apply all the measures which are appropriate for partitions (see above), such as adding a layer of plasterboard, fitted shelving or cupboards.
- Or use your social skills to negotiate a reduced noise level – and don't give up!

Reducing noise from external sources

Noise from traffic and airports can be very wearing. The most important way that this sound is transmitted into the inside of the house is through the windows. To soundproof windows, you require secondary glazing installed with a gap of between 150mm and 200mm between the two panes of glass. Ensuring that all gaps are sealed is another important measure; and you will need to think out your ventilating system very carefully. It could be worth considering a full heat exchange system (see **DRAUGHT-PROOFING AND VENTILATING** on p53).

Pleasurable background sounds

Creating our own pleasant sounds can have a powerful calming effect. If we produce these sounds when we are feeling stressed, we can induce a feeling of well-being and tranquillity. If you live in the country the natural sounds of birds singing, water bubbling or leaves rustling may surround you anyway. However if you live in a town, alternative sources can include a fountain, a hanging arrangement of tinkling shells, or specially produced tapes of natural sounds in order to mask an unpleasant background noise. These can all help to provide a harmonious acoustic environment, reducing stress levels and promoting a sense of well-being.

Priorities for action

1. We can usually do something about all types of noise in our homes. It helps to first identify the source and method of transmission.
2. Be aware of when and how loudly you make noises that might annoy others, such as amplified sound, DIY noise, etc.
3. If you have a problem with noise from the floor of a room in your house, consider introducing a floating floor by the method illustrated (see previous page).
4. If you have a problem of sound being transmitted between adjacent rooms, you will need to increase the weight of the partition, either by adding a layer of plasterboard or introducing integral furniture to add weight and help damp and absorb the sound.
5. Noise from appliances such as washing machines can be reduced by isolating the appliances on rubber feet.
6. If you find the quality of sound in your home a bit dead, or if there is annoying background noise, consider adding your own pleasurable sounds (see above).

Plants

Plants have played a pivotal role in the evolution and maintenance of the planet—regulating the physical environment and providing for the needs of animals, including humankind. We have a symbiotic relationship with plants, and the sooner we re-integrate ourselves with the plant world and work with it, rather than against it, the sooner we will start getting our priorities right. We can begin this process by understanding the many roles that plants have played and continue to play in the evolution of the ecosystem. They embody a natural technology and architecture. The architecture is there for all to see; the sophistication of their technology we are only just beginning to appreciate.

Plants convert carbon dioxide and water into oxygen and organic matter, using solar energy through the process of photosynthesis. Animals and micro-organisms consume most of the oxygen made by the plants, returning carbon dioxide to the air. All the oxygen in the biosphere is produced by plantlife either in the sea or on land. Of course other nutrients are absorbed from the soil and air by plants; and as well as producing cellulose as their basic construction material, they synthesise prodigious numbers of chemical compounds.

Plants can actually provide us with many solutions to problems raised in this book, and in this chapter in particular. As well as providing for our basic needs of food and shelter, they can help us solve many of the problems of pollution that are so urgently in need of ecological solutions. In return we need to learn to look after them and nurture them—a process that will in itself put us back in touch with the natural processes necessary for a sustainable world. However it should be said that house plants are not universally popular: some people dislike the increased humidity that plants engender.

The practical uses of plants in our homes

In the wider world we can see how plants are used to meet our food needs, energy needs and to some extent medical needs, but how can they be used in the home? We shall start with the role that plants can play in pollution control.

Plants as pollution controllers

A NASA scientist conducted experiments to see what use could be made of plants in the confines of a spacecraft. He found that plants were particularly good at cleaning chemical pollutants out of the air. Different plants absorbed different chemicals. Not only that—plants also absorb pollutants from water through their roots, and micro-organisms in the soil absorb pollutants. The most famous example of a plant's ability to absorb a pollutant in this way is the common spider plant, which removes formaldehyde from the air. Formaldehyde is a very common indoor domestic pollutant, originating in chipboard and other manufactured boards and plastics. As an air pollutant it is implicated in 'sick building syndrome' and other environmental conditions. The spider plant will grow almost anywhere—it reproduces more easily than almost any other plant. It is difficult to over-water and thus is a favourite as an indoor houseplant. Keep half a dozen large healthy plants in your kitchen (particularly if it is newly installed), and at least the effects of one indoor chemical will be diminished.

This is just one example. Very little research has yet been carried out to discover which plants can absorb which pollutants, and how they can do this most efficiently. We may find that in the future factories will only be permitted to generate pollution if it can be absorbed by adjacent plantations that can synthesise these toxic byproducts. Solar aquatics, which uses a greenhouse environment to speed up the absorption process in temperate climates, is beginning to provide viable alternatives to much more expensive sewage treatment plants. Countries which have plenty of sun and water have a ready advantage: there is a particularly successful scheme outside Calcutta whereby plants clean the water, and fish which are harvested feed on the plants.

To return to our homes, below is a list of plants that can help clean the air in different ways.

Dwarf banana, *Musa cavendishii*	Spider plants, *Chlorophytum comosum*
Golden pathos	Chinese evergreens, *Aglaonema modestum*
Peace lilies	The *Peperomia* family
Snakeplant, *Sansevieria*	Goosefoot plant, *Nephthytis syngonium*
Ivy arum, *Scindapus aureus*	

There must be many more plants that could be used to solve different pollution problems. One intriguing part of this research is that it has been found that the ability of a particular plant to absorb and metabolise a chemical increased when it was exposed to the chemical for long periods of time. So perhaps plants can learn to use the particular pollutants that are peculiar to your home! Increasing the light levels also increased the rate of absorption. What is essential is that more research is done on the number and size of plants which might deal with particular pollutant loads. It is worth watching out for more detailed information as it becomes available.

Other ways that plants condition the quality of our air

Plants can be used in many ways to improve air quality and provide a more equitable indoor climate. They can help conserve negative ions, they can remove certain odours, and they can moderate the humidity through evapotranspiration in which oxygen and water are discharged into the room.

Besides the plants' leaves, the soil in which they grow can also act as an air filter; it may be possible to design special filter/plant beds through which air is drawn. This method of air cleaning has been tested and shows definite promise. It is particularly the products of combustion that are successfully filtered out.

Plants to clean water

There are various methods that use plants to clean pollutants from water. In this country reedbeds are being experimented with, to deal with household waste water. This is an open air process which centres around a sequence of differently shaped pools or reedbeds through which the water flows. Each pool has a different function, using different plants, microbes and other organisms to extract different pollutants and nutrients. Solar aquatics (already mentioned earlier in this chapter), which uses a sequence of glass containers in a greenhouse, is perhaps the most successful of the many experiments originating in the United States. All these experimental methods are new ways of harnessing biological processes to imitate the way nature cleans dirty water, only carried out in a more systematic manner.

Both these methods produce clean water, processed at a reasonable cost and with useful byproducts. Although not yet developed in a form which could be used in an inner city house, if you live in the country without access to a sewer, the system of reedbeds provides a ready alternative for processing your waste water and sewage (see REFERENCES).

Plants for noise control

Banks of plants can absorb sound very effectively. If you have a particular problem, see whether plants can help solve it. If you experience traffic noise, and have garden space between you and the road, it is possible to grow a dense thicket of trees and bushes to screen the noise. If it is dense enough this thicket will also help absorb some of the car exhaust gases. Inside the home, if you have a reverberation problem, plants can be placed strategically in the room to help to reduce echoes by absorbing sound (see SOUND p.153).

Plants for fragrance

Plants are traditionally used to give fragrance to a garden. Often, however the indoor plants that we use have little smell. The following are a list of those that are particularly fragrant:

The moonflower, *Calonyction aculeatum*
Citrus plants (miniature varieties)

The coffee plant, *Coffea araica*
The tree tomato, *Cyphomandra betacea*
Angels' trumpet, *Datura suaveolens*
Various species of Daphne (*Daphne mezereum* is popular)
Chinese jasmine, *Jasminium polyanthum*
Luculia gratissima
Eucalyptus citriodora

You will have to grow plenty, and place them in the sun, to gain the best results.

Plants for taste and nutrition

To be able to grow indoor plants that we can actually use is particularly satisfying, and helps to increase our motivation. Herbs are an obvious choice. The advantages of fresh herbs over dried ones are important, for some herbs achieve their real flavour only when fresh. Most herbs can be grown inside; however they generally need plenty of sunshine, so potted herbs need to be placed on a sunny window sill or conservatory. Parsley, chives, sweet basil, pot marigold and nasturtium all do well in pots.

Above all, indoor plants can give us aesthetic pleasure. They allow us to bring some of the freshness and beauty of a garden inside the house. By integrating plants into your home and adapting it to their needs, you will find that they will in turn provide for even more of yours.

The external use of plants

Elsewhere in this book, in the EXTERNAL SPACE chapter, plants have been suggested for many different functions:

- as external climate regulators providing shade, windbreaks, insulation, rain protection and UV protection
- as habitats for wildlife and pest-controlling insects
- as barriers for security and privacy
- as coverings for facades and roofs

There remains perhaps the most important aspect of all: how we can look after plants, particularly indoors, so that they can fulfil these functions successfully.

Looking after indoor plants

Plants have certain needs that are very simple. They require water, nutrients, air, light and warmth and perhaps most of all they they need our care.

They will differ greatly in their requirements for each of these, and there are many plant books which outline exactly what you have to do to to provide for and maintain plants in good health. One of the main considerations is the humidity levels that they require. Many houseplants like a humidity of around 60%. Most homes are much dryer than this—many living rooms have a humidity of about 20%. Bathrooms and kitchens tend to have a much higher humidity, of around 50%, and many people choose to keep their plants in these rooms most of the time. Over-watering is a common mistake and it is therefore important that just one person oversees the watering and knows the requirements of any particular plant. The right level of light is also important: different plants have very different requirements. A few plants can survive dark interiors, but most require periodic placing in a sunny spot if they are to remain healthy. Choose appropriate species for different situations in the house, and be aware of the size your plants will grow to. If you get into good habits and are always willing to learn, you are likely to gain much enjoyment and satisfaction from growing and tending for plants.

Bank of plants.

Priorities for action

1. Learn about how to look after plants from house plant books; and learn about the particular needs of the plants that you have.
2. Choose appropriate places in your home for house plants, and work on setting up the right environment.
3. Work out where your most polluting sources are likely to be and place plants close to them.
4. Set up a corner, worktop or cupboard where you can keep all your plant maintenance equipment and books.
5. Set up a little greenhouse, bay window, dormer, or roof light so that you can place plants there to recuperate and also grow new ones from seeds and cuttings.
6. Consider ways of watering your plants with greywater.

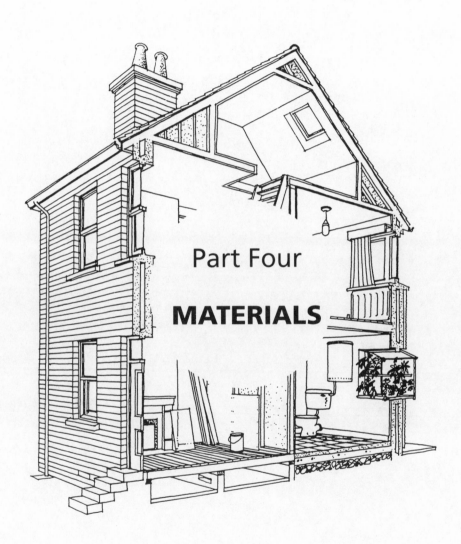

Part Four

MATERIALS

Part Four

MATERIALS

In our climate, building materials are one of the basic needs of life along with food, water and air. However, the way we select and use them has a profound effect on our homes and environment. If we care for the environment we cannot delegate responsibility to a designer, architect or builder for choosing materials unless we can trust them to specify ones that fulfil certain ecological criteria.

On a global basis the amount of raw materials used is staggering. The world manufactures eight times the volume of goods it did in the 1950s—yet this increase has not taken place evenly around the globe. Now that previously low growth regions of the world are beginning to expand production very rapidly, under the influence of the West, the worldwide demand for resources is being stepped up still further. The Far East, especially China with its enormous population, is a case in point. We not only have a population explosion but we have an even greater production and consumption explosion to go with it! The impact on our environment is devastating

In Britain, as in most of the Western world, it is increased personal consumption that is the main cause of this increased volume of production. We live in a growth economy where there is always the underlying assumption that it is good to increase the size and quantity of personal acquisitions, and improve their quality. There are many examples of these continually increased expectations: from refits for kitchens, bedrooms and bathrooms to ever more cars, demanding ever more road building.

What is worse, we now use materials in far more wasteful ways than ever before. Wastage on building sites often amounts to about 20%; and there are the mountains of refuse that come out of our homes every week. It is only our short-lived energy glut and unsustainable growth that are allowing us to waste on this massive scale.

There is much we can do to turn this tide. The rise of green consumerism has shown that together we have an economic power to affect ecological change. First we need to become well informed on the criteria for choosing particular materials or appliances. We need the will and determination to make decisions independent of 'the Jones's' and of TV advertising. We need to know when not to buy but to clean up and repair whatever it was that we were going to replace. However in order for us to be able to behave in an ecologically responsible way, we also need accurate and relevant information from manufacturers.

To provide information, manufacturers also need the right priorities and the incentives to produce ecologically. We ourselves can provide those incentives by supporting those companies heading in the right direction. Some progressive businesses are realising that they have little choice in the long run but to produce products that are acceptable environmentally. We have to create a climate in which companies will not succeed if they have a careless approach to the environment.

New ecological standards are required. There are clear national and international standards in terms of strength, safety, durability etc. regarding the quality of particular products. However, standards and information relating to environmental impact are more difficult to measure and are rarely given as part of product information. Some green labelling systems got off to a bad start. Now the EC is starting a new labelling system, with each country developing standards for different materials and appliances. We await the results of this new initiative with interest. However we need never wait for governments to act before acting ourselves.

The first chapter starts with a look at the different criteria that we can apply to help us choose between different materials. The materials themselves are then analysed, starting with the renewable materials, which are taken mainly from the plant world, followed by mineral resources and ending with a look at synthetic and processed materials.

The consumer revolution has had many effects, most of them detrimental to the environment. Now we need a greening of this consumerism—and, most of all, less consumption.

Criteria for materials selection

What criteria should be applied when selecting materials for incorporation into an ecological design or renovation? This chapter aims to answer this question by briefly elaborating on ten chosen attributes. Ecological building materials are those that comply with most or all of these which are listed below. In many cases it will be a matter of finding a material that provides the best balance between the ten different criteria.

- *Clean*—or non-polluting in terms of:
  - global warming
  - ozone depletion
  - acid rain
  - ground and water pollution
- *Healthy*—(to humans and domestic animals) i.e. most commonly natural or inert materials
- *Renewable*—most usually products of living organisms, such as trees
- *Abundant*—of almost inexhaustible supply, for instance certain rocks
- *Natural*—as highly processed or synthetic materials tend to waste more energy and raw materials
- *Recyclable*—reusable or biodegradable
- *Energy-efficient*—low embodied energy consumption in production
- *Locally obtained*—and often part of the vernacular building traditions
- *Durable*—reducing the need for frequent replacement
- *Design-efficient*—favourable for flexible, safe and efficient design

Clean or non-polluting

In this category, non-polluting refers to those materials which cause a minimum damage to the earth's ecosystem. The materials *excluded* from this category are those that cause atmospheric pollution, in particular those gases and vapours causing global warming, ozone depletion and acid rain. In addition to atmospheric pollution there are those materials which cause the pollution of the ground, rivers, lakes and oceans. Each of these different types of pollution are briefly analysed below.

Greenhouse gases

Greenhouse gases have already been referred to in the ENERGY chapters. The most important are carbon dioxide, methane, nitrous oxide and the CFCs. Materials involving CFCs are looked at under ozone-depleting chemicals below.

Most materials are responsible for some carbon dioxide emissions; it is possible to evaluate all materials in terms of tons of carbon dioxide produced per ton of material. It is then possible to calculate the total amount of carbon dioxide produced per building component or appliance. There is a rough correspondence with embodied energy (see ENERGY-EFFICIENT section below).

Methane is now being taken more seriously as a greenhouse gas. It is caused by a number of factors which mainly relate to agricultural practices. However a growing proportion of methane is given off from garbage dumps where anaerobic decomposition takes place. This gives us a further reason to be more aware of what happens to our household waste, including our sewage, when it leaves our home.

Nitrous oxide (N_2O) pollution is caused by nitrogen-based fertilisers, fossil fuels and the burning of biomass. It is thus mainly an energy use issue, in terms of building materials (see ENERGY-EFFICIENT section).

Ozone-depleting materials: CFCs

Here we shall simply take on trust that ozone depletion is a serious threat to our environment, and its ultimate consequences are as yet unpredictable. Besides causing ozone depletion, the CFC-12 molecule is 10,000 times more damaging as a global warming agent than a CO_2 molecule. On the basis of these two effects it seems sensible to eliminate CFCs as completely as possible from further use.

There are four ways that CFCs are commonly used in domestic building:
- As the expanding gas in insulants such as polyurethane foam. This could be replaced in many cases by carbon dioxide, helium or argon.
- In refrigerators, where the refrigerants can now be replaced by a mixture of propane and butane. Refrigerators using this mixture are now on the market.
- As propellants in aerosol sprays such as paint or foam; carbon dioxide, helium, argon and even propane in certain circumstances can be used instead.
- In fire protection equipment (halon), which can mainly be replaced by carbon dioxide.

Since there are these perfectly reasonable alternatives, there is no need to buy building materials that use CFCs.

Acid rain-causing chemicals

There is some uncertainty as to whether coal-burning power stations or the agricultural use of sulphates are more to blame for acid rain. Whichever it is, it seems sensible to reduce the pollution from both these sources. The following table gives the acids and their main man-made causes.

- Sulphuric acid, produced by sulphur oxide emissions from power stations and fertilisers being broken down in the sea
- Nitric acid, produced by nitrogen oxides in combustion gases
- Formic acid, produced by methane oxidation (see methane above)

As can be seen from the above, in terms of reducing acid rain that can be ascribed to the use of building materials, we need to be aware which building materials produce quantities of these gases either directly (as in brickmaking or cement production) or indirectly through high-energy processes (see **ENERGY–EFFICIENT** section).

Ground and water-polluting materials

Products which produce ground contamination or degradation of the earth's skin include those involving the destructive extraction of raw materials, including deforestation. In many cases ground water is polluted in the process.

In the **WATER** chapter, mention has been made of some of the most important water-polluting chemicals. Although agriculture is the biggest polluter of water, through pesticides and fertilisers, industrial processes run a close second and it is here that we need to know a lot more about the water-polluting nature of different processes. Nearly every industrial process uses water at some point: it is a matter of how much water is used and the total amount of pollutant that is leaked into the environment (into ground water, river, lake or ocean). Many paints, for instance, are water-polluting, such as cadmium which has largely replaced lead as the pigment in brilliant white paint.

Healthy

This subsection also deals with pollution, but of a different type. Here our concern is for personal health rather than the health of the planet. The main issues have been studied in the early chapters of **HEALTH** . However, it might be useful to understand the difference between a hazard and a risk, in relation to health. Lead, for instance, is a *hazardous* material, but the risk of it proving a danger to our health depends on where it is, the quantity of it, and the likelihood of it contaminating our food etc. It is therefore possible to have a small quantity of a very hazardous material in a safe and secure location posing very little risk, whereas a much less harmful material which is everywhere in the house, such as PVC, a constituent of most paint and many plastic household products, could cause a greater threat. If in doubt, use an obviously benign material if one is available. For instance if you are buying a carpet and have the choice, it is safest to buy one made from natural materials such as cotton, jute and wool.

On dangerous chemical containers there should be a 'Hazardous products' safety notice. Take these notices seriously and read them carefully. Only use such products if there is really no alternative.

Besides diet, exercise and lifestyle, the most important health influences encountered in the home are the materials that are in contact with our bodies, and those found in the air we breathe. Contaminants from building materials in the home that affect personal health include asbestos, organochlorines, dust from treated timber, formaldehyde, lead, phenols and volatile solvents; and also the products of combustion, household cleaning, maintenance and pest control (see TOXINS on p.115).

Renewable

The only materials that are truly renewable are the products of living organisms that use energy directly or indirectly from the sun, and are made up from compounds that are continuously recycled in the biosphere. Although there are some such materials that are the products of animals, such as leather, silk and certain glues, the main source of these is the plant kingdom.

It is also generally accepted that labour can be seen as a renewable resource. This is one of the features of ecological thinking that results in very different analyses, when compared with conventional economic thinking.

There is an important difference between fast-growing plants that provide easily renewable materials and those that are rare or slow-growing. A renewable source should feature a harvesting cycle related to its replacement growth. Trees that take a hundred years to grow need to be harvested in a cycle of similar timescale!

All stone, sand, aggregates and clay are renewable, but on a geological time scale.

Abundant

One of the most important features of an ecological building material is that there should be abundant supplies of it. We can of course have an influence over which living resources are abundant enough for us to harvest: we already manage many deciduous species of softwood, and increasingly hardwood, by planting them on a sustainable basis. Increasingly there is the idea of mixing species within a plantation to provide a healthier environment, thus producing greater abundance (see LIVING RESOURCES chapter).

Of our mineral resources there are some that are rare and some that are abundant. Stone in many of its forms is the obvious example of a very abundant material. The problem here is how to extract stone in a way that is least damaging to the environment. Problems occur when the wrong site for extraction is chosen – a national park for instance – and the volume of stone removed by the extracting company is out of proportion to the scale of the landscape.

Natural or unprocessed

All materials require processing or shaping to some extent before they can be used. Those that are synthetic or require very heavy processing are much less likely to be ecologically sound. Though not a hard and fast rule it should, however, make us think twice before accepting a processed material when an unprocessed one might be more appropriate.

Examples of highly processed materials are most of the metals, all plastics and, to a slightly lesser extent, cement and glass. The last example is a material which has so many other positive ecological attributes (energy saving, healthy, light transparent, strong, durable, non-polluting, recyclable, abundant, etc.) that its high processing costs are balanced by these considerable advantages.

There are also an enormous number of synthetic, organic compounds that do not occur in nature. Many of these are hazardous to our health, to a greater or lesser extent. Some of course have been particularly developed to be destructive to life, such as the biocides. The main reason that they pose such a threat is that they are produced industrially on such a massive scale. The word organic here is almost perverse as it merely means a substance containing carbon, rather than its original and commonly used meaning of being derived from living organisms. However most organic chemicals are derived from the processing of oil or coal, which have themselves derived from living matter (see **PROCESSED AND SYNTHETIC MATERIALS** chapter on p.194 for further information).

Recyclable

Recycling is one of the fundamental principles of ecology. Every item of waste is a potential input to another use or process. If a by-product is produced which cannot be used, is toxic and difficult to reprocess, it is questionable whether the by-product should have been produced in the first place. We need to think of all the resources we use as part of much more sophisticated and interdependent cyclical systems.

Waste is a new phenomenon which has been accepted as quite normal— yet until two centuries ago there was virtually no waste, and what waste there was, was biodegradable. At present the quantities of waste that are produced by our society are truly staggering. The building industry provides a terrible example. We have the idea that we can throw something away, but we are learning to our cost that the dustbin is our very own earth (see **RECYCLING** chapter on p.205).

Energy-efficient

Energy costs can sometimes be a very high proportion of the total cost of materials. Around 10% of all the UK industry's energy requirement (or 5% of the total UK energy consumption) is used to produce and manufacture building materials and products. This energy is now referred to as 'embodied' energy. Most of the materials that embody this energy are used in the renovation and extension of existing buildings because the annual rate of new building is only a very small proportion of the total existing stock. In the housing sector the annual new-build rate is at present no more than 1% of the total housing stock.

In any particular house renovation most of the materials, and thus the embodied energy, are already in place. However, given that domestic improvement cycles are often as frequent as every 10 to 15 years, the embodied energy value of improvements can compare with the energy used in occupying the houses. In the past houses were not redecorated so often; nor were kitchens, bedrooms and bathrooms constantly refitted.

Life cycle energy costs

What are the different ways in which energy is consumed in the life of a particular building material, whether a part of new-build work or renovation?

Energy used in
  • its extraction
  • any processing or manufacture
  • transportation
  • its installation
  • its maintenance
  • its demolition or recycling

The most relevant and enlightening comparisons can be made in the energy costs of production: the energy consumed in felling, sawing, and transporting timber has been estimated at 580 watts per tonne. Taking this as a baseline, the energy costs for some other materials can be approximately measured and compared as followed:

Aluminium: 126 times the average timber processing energy
Steel: 24 times
Glass: 14 times
Plastic: 6 times
Cement: 5 times
Bricks: 4 times

Of course these figures need to be interpreted in terms of the amount of the material required to fulfil a particular function. Also it is important to note that 90% of the energy needed to make wood comes from the sun, whereas all the energy needed to make a fired brick comes from fossil fuels.

Locally obtained—vernacular

There are a number of reasons why locally obtained materials are an ecological choice. The cost of transport is an obvious consideration; however, the use of the local vernacular building materials is perhaps the most interesting. In the past, vernacular construction methods and styles developed from the use of local materials that were readily available. The oldest vernacular building construction used wood and mud, and later thatch for roofing. Gradually the materials that were found to be most durable and efficient were used in preference, and styles that were appropriate for the local climate were developed.

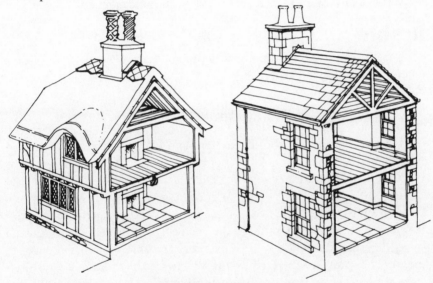

Vernacular use of materials.

Over time, regions developed styles of building and crafts that were harmonious and appropriate to particular localities. It was only with the coming of extended transport systems that house-building materials from other regions became available. Fashion and ideas imported from abroad then began to take precedence. Sometimes this led to an improvement in the vernacular style, but it could equally well lead to particular images that people wanted to emulate. Vernacular styles were then looked down on as not being fashionable, or else they were associated with the oldest and often poorest housing. National standards often destroyed local styles and traditions. Developers had little idea of keeping to the vernacular. Finally, the increased movements of populations led to the loss of an awareness of distinct local styles.

In other words, if the green philosophy of building for durability is to be generally adopted, it is likely to result in much greater interest in localised vernacular design at every level—from the general approach to scale and orientation to details of external joinery, roofing and walling materials and actual methods of construction.

We must develop a local vernacular again. This will give each locality a unique style, which is exactly what we like about attractive areas such as the Cotswold or Dales villages. There is a lot to be said for each area once again developing its own building materials and styles, developing a particular character and employing local craftspeople.

When renovating your house—if it is an old one—it is useful to identify the original features and materials that were used and to consider which of these could be reinstated. It is important to be able to distinguish between mock vernacular—often planted on—and good local practice.

Durable

There are different types of durability, which include
- Strength or non-breakability
- Non-inflammability
- Ability to withstand water or high humidity
- Ability to withstand attack by pests or fungus
- Ability to withstand exposure to light and UV radiation
- Ability to withstand attack by chemicals
- Endurance against mechanical or electrical breakdown

It is important to see that we need to make things appropriately durable: for instance, a polythene bag floating in the sea is inappropriately durable, as are washing detergents that are not easily biodegradable. When it comes to elements of our houses we need the basic structure to be as durable as possible, whereas those parts we want to change from time to time, like the wallpaper and certain paint surfaces, need to be biodegradable at the right time. Worthwhile things like a well-crafted chair or table should be durable and, perhaps even more importantly, the windows and doors. If we look at successful design that has lasted, we are often given a clue as to which materials and design achieve this—leading us to the final criteria of design efficiency.

Design-efficient

Although this is not strictly a characteristic or quality of a particular building material, it is of such importance that it needs to be addressed. What might some of the criteria of design efficiency be? The following are eight considerations to be taken into account, in addition to the ones above, when making a choice of materials:
- Minimum use of materials—producing small and light components

where appropriate, and miniaturisation in some cases.

- Simplification of design—to fulfil its basic function.
- Safe design for fire—besides being life-threatening, fire is an indiscriminate destroyer of household resources.
- Multifunctional design—leading to versatile products.
- Design from natural forms—if we look carefully at what has succeeded in nature we see just how well adapted and designed living structures are.
- Design to shed water—and prevent damp and decay.
- Design to prevent insect attack—and resultant impairment.
- Design for solar efficiency—to maximise the use of the sun's energy.

Good design is an integral part of the use of ecological building materials. Now it is up to us to have some fun finding ways to achieve this!

Living resources

When, in the last chapter, we looked at what constituted a renewable material, it seemed that life itself was the essential element. Life of course can only exist with the help of minerals on the surface of the earth and energy from the sun; however, both these are relatively permanent compared to most life forms. It seems then that life itself remains the most vulnerable and vital ingredient that lies behind the secret of sustainability. This chapter looks at those plant and animal products we use as materials in our homes, and how we can harvest them ecologically.

We should also remember to treat with respect all materials that have been part of living organisms. Many of us have become so alienated from life outside our towns and cities that we easily forget that we rely totally on our living ecosystem. What is also thought-provoking is that our bodies are made up of atoms and molecules that have been part of many previous plants and animals, and are in constant interchange with the air, water, minerals, plants and animals about us. We should not see ourselves as separate from this ecosystem—we are an integral part of it.

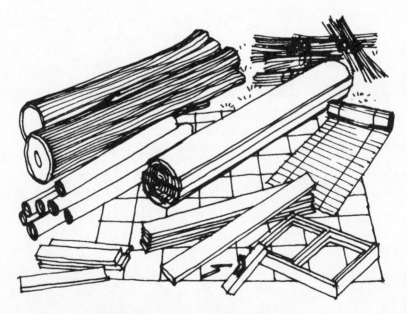

Products from living resources.

Considering all the products from living organisms, timber and its products are by far the most important for our housing; an analysis of the way timber is treated forms the major part of this chapter. We should not forget though the other plant and animal products—such as reeds, canes, kapok and wool, which are briefly discussed towards the end of this chapter.

Timber

Timber is the ultimate ecological building material just so long as it is managed properly. It can be used for almost any part or element of the home. It is also one of the healthiest building materials and can be used as a natural regulator of indoor climate, helping to stabilise humidity. It is warm to touch, absorbs sound and remains unaffected by electromagnetism. Timber is also a natural insulator, which gives it natural energy-saving characteristics. It requires a smaller quantity of energy to process than most materials, involving very little waste or pollution. It is recyclable and biodegradable, has a natural durability if kept dry and a high weight to strength ratio. It is also one of the most attractive building materials, coming in a wide range of colours and textures. Surely hard to beat!

If we are to use timber sensibly in home renovation, we need to understand the principles of sustainability, the destructive forces that are at present destroying the rainforests, our role in encouraging this destruction and how to counteract this by knowing which timbers to avoid buying, particularly the tropical hardwoods. We also need to be sure that the timber and timber products we use are from sustainable sources. And finally we need to know how we can preserve the timber in our homes without the use of damaging chemicals.

The destruction of the rainforests

The destruction of rainforests is one of the most urgent ecological problems facing the world at the present time. Some people would rate it the most serious problem of all. There is a common misconception that the rainforests provide additional oxygen for the world, but this is largely untrue since nearly all the oxygen produced by the trees is taken up by the animals and microorganisms within the rainforest. As well as being valuable in themselves, it is as an air-conditioning and climatic control system for the planet that rainforests are particularly important. They also protect the soil from erosion and provide a habitat for innumerable species of life.

Why is the wholesale destruction of the rainforests continuing at an ever increasing rate? Vested interests play an important part, although the timber loggers and exporters are dependent on importers, distributors, retailers and ultimately on us as customers.

In fact, a study carried out by the Oxford Forestry Institute has estimated that 90% of revenues from the products of timber from the rainforests go to the individuals, companies and agencies in the importing counties. Less than

1% goes to the government of the exporting country. The result is that the governments of some tropical countries may receive less in timber revenues than it actually costs to run their forestry services. The worst affected and most vulnerable are the tropical rainforests of Africa, Asia, Malaysia and South America.

Besides the logging of tropical rainforests, there are many other ways that these forests are under threat. Felling and burning of trees to make way for cattle ranches, roads and short term smallholdings are all taking their toll. In temperate climates, acid rain is affecting more than a quarter of all trees and the pressures on land for development are often heavy. However, perhaps the greatest threat comes from our insatiable demand for paper. Recycling is not a comprehensive solution as paper is not as easy to recycle as many think— the ink needs to be washed out and the quality of the fibres is reduced in the process. In many parts of the world timber has been the traditional fuel, but population pressures and desertification are combining to denude even impoverished landscapes of their greenery. And with the changes in climate due to global warming, forest fires are becoming an increasing threat in some areas.

Sustainable forest management

In Britain there are of course many forest management practices that are less than beneficial, such as wholesale clearing and replanting. Sustainable multi-use management of deciduous forests provides an ecological way of managing the renewal of timber and other resources. From the agricultural point of view, there is increasing interest in forest farming and there is now the real-isation that this also makes sense from the viewpoint of good forestry. Areas of forest can be zoned for different activities and experiments carried out using different mixes of species. Animal life becomes an integral part of the scheme by providing habitats for species that need to be encouraged, as well as sanctuaries for wildlife under threat. Timber harvesting in the multi-use forest requires a careful balancing of rates of extraction, including some benign neglect. Disturbance is kept to a minimum. All we are doing is allowing nature to act in the most efficient way she knows and this requires us to understand better the way she works and get in the way as little as possible. Over-exploitation is usually prevented by only cutting trees of particular diameters. However more sophisticated ecological methods of management will be developed as understanding of the natural cycles grows, and we learn from the understanding and experience of native peoples who have lived for centuries in harmony with their forest environment.

At present there are very few forests in the world that are being managed on these principles, and many forestry managers remain ignorant of their potential benefits. The re-establishment of trees in areas of devastation is one way to heal and revitalise areas which would otherwise be barren. This can make an immeasurable contribution to redeeming the environment, and offer an ideal use for areas where industry or agriculture are not present or not desirable.

How to avoid buying unsustainable timbers

Since a recent study for the International Tropical Timber Organisation (ITTO) found that less than 0.2% of tropical forests are being managed sustainably, it may make sense to avoid using tropical hardwoods altogether until it is clear that the problems of sustainability have been overcome.

However if you are determined to use tropical hardwoods for a particular premium purpose, ensure that the source of timber is certified as being the product of managed, sustainable forestry by the relevant government agency in the country of origin. It may be very difficult to be sure of what you are being sold, so it is worth reading around the subject with the help of the many guides available such as *The Good Wood Guide* (see **REFERENCES**). Pressure from knowledgeable consumers may help as much as a total boycott.

In most cases—and this applies to virtually all building joinery, such as window frames, staircases and standard doors—substitution of temperate hardwoods and softwoods is always possible. It is very easy to be sold tropical hardwoods as a component in plywoods or other types of boards for building purposes. This is an appalling waste of valuable timber, which should be conserved for special uses. Many UK board importers and suppliers are now making an effort to improve the quality of plywoods made from temperate sources.

Tropical hardwoods we can use

In most cases we should avoid using tropical hardwoods. However, there are a limited number which are from sustainable sources and whose use should be encouraged in preference to other timbers:

Timber	Source
Greenheart	Guyana
Rubberwood	Malaysia
Teak	Some sustainably managed forests now exist in Java, Thailand and Burma

Temperate hardwoods we can use

The following list details woods with characteristics which make them ideal as replacements for tropical hardwoods. Some are especially appropriate for externally exposed joinery such as window frames, and others for use in internal joinery. Native hardwoods often have a greater inherent resistance to pests and weathering than many of the exotic timbers now widely used. The timbers listed below are increasingly available at competitive prices and in commercial quantities.

Timber	Uses
Alder	Plywood and furniture
Ash	Fittings and furniture
Beech	Furniture and flooring
Birch	Plywood, wood components and flooring
Cherry	Special items of furniture
Chestnut	Excellent for outdoor use
Elm	Structural use (water–resistant)
Lime	Easily worked for carving
Oak	Durable for external joinery, window frames etc.
Pear	Carving and small special items
Plane	Veneers and furnishings
Poplar	Internal carcassing and flooring
Sycamore/maple	Internal use, carcassing and with harder types; flooring.
Walnut	Decorative uses, veneers
Willow	Flooring, carcassing and fencing

It is important to understand the properties of these different timbers: there are many books which give details of what to expect in terms of durability, stability, workability and the likelihood of splitting. The more the pitfalls of any particular timber are understood, the more it is possible to find ways of overcoming these through correct detailing and design. It is particularly important to know how to design for protection with external use.

Timber products: efficiency and use of waste timber

There are many products that use timber as the main material:
- Veneers
- Blockboard
- Plywood
- Chipboard
- Medium density fibreboard (MDF)
- Hardboard
- Fibre insulating board
- Woodwool

At present many of these contain tropical hardwoods and should be avoided. Some manufacturers are now returning to the use of indigenous regional timbers as demand for products made from these sources increases. See **REFERENCES** for help in locating manufacturers, products and suppliers.

Formaldehyde glue is used in many of these products, including chipboard. With products containing formaldehyde the main consideration is where the board is to be used. If it is indoors in a badly ventilated space or in a bedroom, it would be worth finding an alternative; formaldehyde-free materials are beginning to be easier to find and identify. However there is always the choice of using these products and growing spiderplants to absorb the formaldehyde vapour (see **PLANTS** in **HEALTH** chapters).

Fibre boards such as hardboard, medium-density fibreboard and insulation board are made of materials ranging from wood pulp to shredded sugar cane. Some of these are made in a similar way to paper without the use of glues or synthetic resins. If you are keen to avoid products containing synthetic resins and formaldehyde it may well be necessary to check with the manufacturers to find out how they are made.

Linoleum

Linoleum is a material which is coming back into use again, PVC having largely taken over its role in recent years. Linoleum is made of entirely natural products: powdered cork, linseed oil, wood resin and wood flour, with the material being pressed onto a backing of hessian or jute. It is flexible and strong and comes in sheets or tiles. It feels warmer on the feet than PVC and is 'softer' in feel. It should be installed on a smooth, firm and damp-proof surface with a lignin paste.

Cork

This is a renewable material, and environmentally benign. It is the bark of the evergreen oak, *Quercus suber*, and the stripping of the cork bark surprisingly does no harm to the tree; in ten years it is ready to be stripped again. The cells of the bark are filled with air and can withstand very high pressures without rupturing, returning to their former size when the pressure is withdrawn. Cork is light-weight, durable, non–flammable and excellent for flooring, insulation and veneers. It is produced on the Iberian peninsular by small firms and its use should be encouraged.

Rubber

Natural rubber comes from latex extracted from the rubber tree. In most cases its uses have been replaced by synthetic rubber and plastic. It is still possible to obtain natural rubber floor tiles and natural rubber foam for furnishings. This is well worth doing as the properties of natural latex in terms of durability and flexibility are superior in many ways to those of its synthetic substitutes. It also burns less easily and less toxically than many of the plastic foams used in furnishing. Oxidation is its main cause of degradation; protection from this process prolongs its life. It is also used in adhesives, and to form resilient layers for underlay and soundproofing.

Paper

The use of paper in newspapers, magazines and office stationery results in the development of vast conifer forests purely for this purpose. These forests are often ecologically barren, causing soil erosion and heightening the acidity of rivers and lakes. The process of papermaking itself can also be very polluting. This having been said, our insatiable appetite for paper lies mainly outside its use in the fabric of our homes, and its use in building should generally be encouraged.

In building, the main uses of paper are for wallpaper and plasterboard. Ecologically, wallpaper makes a wall finish preferable to most paints, which contain many toxic chemicals and cannot easily be recycled. When wallpaper is finished with, it can be stripped off and will biodegrade satisfactorily. This biodegradability is threatened, however, if thick layers of toxic synthetic colours are used.

Insulation from recycled newspapers

The use of processed waste paper as an insulation material is growing quickly in popularity. It has insulating properties superior to fibreglass and mineral wool, its main competitors, if it is properly protected from damp. It is treated with borax to make it fire- and insect-resistant. When installing this material there is a choice of placing it by hand or spray-blowing it into place.

Avoiding timber decay

Traditionally, the control of decay in timber has been achieved through proper seasoning and careful detailing. However, more recently, there has been more and more reliance by the timber trade and builders on the use of timber-preserving treatments in order to control rot and infestation. This has led to a general slackening of good seasoning practices and design that protect timber naturally. At the same time there has been an increase in the sealing of buildings for energy conservation purposes, which has often increased the risk of condensation and reduced the ventilation that timbers need. And since we are now more aware of the toxic effects of many wood-preserving chemicals, we need to return to using the best traditional ways of controlling decay as well as developing new ones.

To season timber properly the bark must be stripped and the log soaked in water to remove the sugars in the sap. These sugars can be a major factor in helping rot to gain a foothold. The soaking process, which so often is ignored, can be achieved either by floating logs in a river or pool or, alternatively, spraying them with water. Kiln drying speeds up the process of drying out timber, but for some timbers does not give them time to stabilise. This not only wastes energy but results in subsequent warping and shrinkage, leading to further waste.

With renovation work, we have two situations to deal with. First, any new work involving timber requires to be appropriately detailed to avoid future decay; and secondly the most ecological treatment of any decay that is found needs to be identified. In the case of new work, it is important to choose well seasoned timber and to install it in such a way as to ensure that it will remain reasonably dry. The best way of doing this is to provide ventilation around the timber if it is situated near a possible moisture source. Most of the traditional methods of building adhered to this principle by providing ventilation both in the roof space of the house and the underfloor where there is most risk from either rain penetration or rising damp.

When treating rot in the existing structure, it is important to identify the cause of the decay. In most cases this will be a source of moisture combined with a lack of ventilation, often due to the sealing of ventilation routes or other means of moisture escape. By reducing any source of moisture, drying out the timber and increasing ventilation, the problem should be solved so that there should be no need for toxic timber treatment—which can often prove worthless in the long run anyway. In difficult cases it may be worth obtaining specialised help.

Other plant products

Besides timber there are many natural, renewable products from plants and animals. These include materials from thatch to silk. A brief resumé is presented of each material and its properties.

Canes and grasses

These have a relaxing quality about them. They are lightweight and have insulating and acoustic properties. Bamboo in particular has been used very successfully for all manner of building functions. Thatch of course has been used in this country for centuries. Now, however, some of these reeds are becoming the last remaining habitats of certain species of birds and animals. Such canes and grasses need to be cultivated in carefully managed plantations and fields.

Straw

This has long been a traditional material for roofing. However it is not long-lasting and the mechanical methods of harvesting break up the straw. Norfolk reeds make a more appropriate thatching material (see below). Straw can be made into strawboard, used for roof decking and strawbales are now being used to construct small scale superinsulated buildings.

Reeds

Reeds are much stronger than straw and are now the preferred material for thatching. The Norfolk reed is the best in Britain, growing to a height of about 2.5 metres. These reeds can last as long as 50—60 years. However it has been found that the use of nitrates weakens the durability of reeds, so organic sources need to be used for thatching. Reeds are also used for many types of matting. Such mats don't last so well if they are kept in too dry an atmosphere, but if sprayed with water occasionally their durability is increased.

Bamboo

This is another perennial grass which is not grown much in Britain except ornamentally. There would seem to be great opportunity to develop species that grow well in temperate climates (there are over a thousand species in all). Its very high strength to weight ratio makes it ideal for furniture. Wall panels, woven screens and floor coverings are among many other uses.

Fabrics and fibres

Although cotton is widely used, many of the other natural fibres are going out of use because of the growth of synthetic substitutes. We need to rediscover the aesthetic and practical advantages of some of these materials.

Linen

Linen is an extremely strong natural fibre whose strength increases when wet. Linen made from flax has become expensive because its processing involves more elaborate processing than cotton. However it is long-lasting and can be grown locally.

Cotton

Produced from the cellulose fibres of the cotton seed pod, cotton is the most versatile of the natural fibres. It can be woven into a wide range of fabrics from thin muslin to tough canvas. The ecological impact of intensive cotton growing can be considerable where there are very high chemical applications of pesticides and fertilisers. Also, irrigation demands can upset traditional water cycles, as has happened around the Aral Sea. Many cotton materials are also highly treated with chemicals to give them different characteristics. For instance urea formaldehyde is used to give cotton a non-iron quality. The benefits of the use of cotton, as with the use of so many natural materials, are thus clouded by these considerations.

Kapok

The silky fibres from the silk-cotton tree can be used for the fillings of bedding and upholstery; however, when worn, the fibres break down to produce a dust to which some people are allergic.

Jute

Jute fibres are traditionally used for making hessian for use in sacking and wall coverings. Its main use in our homes is likely to be as the backing for linoleum and carpets.

Hemp

The hemp plant produces strong coarse fibres used for cord, rope, matting and cloth. The main problem is that the growing of hemp or cannabis is strictly controlled because of its use as a drug. It also produces high quality paper, owing to its long fibres.

Sisal

Sisal fibres, from the American aloe plant, are exceptionally strong and can be made into cord and matting.

Coir

Coir comes from the coarse fibres of the outer protective husk of the coconut and is mainly used for door matting.

Animal products

There are three main animal products in wide use—wool, silk and leather. At present there is a glut of wool due to its substitution by synthetics. However it is a strong, durable material and is almost certainly the most appropriate material for carpeting, ideally with hessian backing. It wears well and is healthier in the home than synthetic alternatives.

Silk is spun from the cocoons of a moth which no longer exists in the wild. It has an extremely strong weight to strength ratio and can be dyed most effectively into brilliant colours. It is used for special furnishing where a rich quality is desired.

Leather is used as furnishing material and is extremely hard wearing if it is looked after.

Priorities for action

1. Choose sustainable sources of timber.
2. Avoid tropical hardwoods except those mentioned on p.179.
3. Become aquainted with indigenous hardwoods and use them where appropriate.
4. Use timber composite boards such as ply and particle boards where this can result in more efficient use of timber.
5. Also consider using
   * other timber products such as linoleum, cork, rubber, wallpaper and cellulose insulation from newspapers
   * grasses such as reeds, straw and bamboo
   * natural fibres, from linen to coir
   * animal products—wool, silk and leather

Mineral Resources

The Earth is a sphere of minerals 12,700 kilometres in diameter, molten in the outer core with a plastic solid mantle and thin brittle crust. At present we can only be interested in the crust in terms of mineral extraction, and particularly the surface of that crust where so much of land life is concentrated. This layer of topsoil, vegetation and animal life can be seen as the all-important living skin of the planet. It is this vital surface layer that we are destroying at an alarming rate through deforestation, agricultural impoverishment, erosion, desertification, urbanisation and paving for roads. We are also destroying and polluting parts of this layer through careless extraction of minerals.

Besides this surface scarring, what should our other ecological concerns be in regard to mineral extraction? They are the energy and resources used in the processes of extraction and restoration, the interference with groundwater supplies, the speed of resource depletion and, finally, the pollution caused during the refining process. This last concern is addressed in the next chapter, which looks at **PROCESSED AND SYNTHETIC MATERIALS**.

Perhaps the most important of all these concerns, which we continue to ignore, is the over-mining of scarce resources for short-term gain. Although the known reserves of particular minerals are continually being extended, we are extracting first the deposits that are easiest to mine, so there is likely to be greater and greater environmental damage as we continue to extract them.

Categories of minerals

There are many different ways in which minerals can be classified; one interesting category from an ecological point of view is that of mineral deposits that are a product of dead organisms. These include the fossil fuels of coal, oil and natural gas, the calcium deposits of chalk, limestone and marble, and the uranium deposits of yellowcake.

There are more than 2,500 mineral species. Rocks, which are composed of minerals, are often referred to generically as 'minerals' by industry. Industry at present depends on some 80 important ones, including assemblies of minerals such as iron ore and bauxite (aluminium ore), which are relatively plentiful.

For our purposes, it is useful to divide minerals into those that can be used in their natural state without further processing, such as building stone, and those that are the raw materials for further industrial processing, such as the metallic ores.

Products from mineral resources.

Stone

Stone is a generally abundant but non-renewable material which is expensive when compared with concrete or brick. One of the main reasons for its expense is the labour that is required for both its extraction and incorporation into a building. Most building stones are strong, durable and attractive, which has often made it the first choice as a building material for the wealthy. It is also generally a healthy and non-polluting material. Its embodied energy costs are variable, depending on the ease of quarrying and transportation, with the latter becoming particularly high if the stone is imported from abroad.

Extraction

If extracted in an ecologically sound manner, there is no reason why stone should not be fully exploited. The problem here is how to extract it in a way that does the least damage. With creative stone extraction in derelict or uninteresting countryside followed by careful landscaping and restoration, an area can even be improved—as opposed to the clumsy and destructive quarrying that is often the case today. Problems occur when an inappropriate site for extraction is chosen—a national park for instance—and when the extracting company is allowed to remove too large a quantity from the same place. Experimentation is taking place with new methods, one of which is the 'glory hole' method where stone (often granite) is removed from the inside of a giant hole within a large hill or mountain. Access is obtained from the bottom of the hole, and the top is secured for safety.

Different types of stone

Geologically, stone may be classified into igneous, sedimentary and metamorphic. *Igneous* is rock cooled from molten magma; it often has a coarse crystalline structure (for example granite). *Sedimentary* rock was formed at low temperatures, generally as sediments at the bottom of the seas, oceans and lakes. Typical examples are shales, sandstones and limestones. *Metamorphic* rock has been changed through great heat and pressure. Examples of this type of rock are: slate (changed from shale) and marble, which began as limestone (a sedimentary rock) and became marble through increased pressure and heat. There are in Britain many kinds of building stones, which mainly fall into five types: granites, sandstones, limestones, flints and slates.

Granite

Granite is the strongest of building stones found in the British Isles. It is very resistant to weathering and acid rain, and it can be recycled almost indefinitely. It is used as building stone where permanence is required, and in sets and kerbstones where its strength and wear resistance comes into play. It is used extensively as chippings in the wearing layer of roads and motor ways. Some granite contains fairly high quantities of uranium and thus gives off radon gas at a greater rate than other stone. If you live in a building made of or built on granite you can have your radon levels checked to see if any protective measures are required (see **RADON** on p.142).

Sandstone

Sandstone is a sedimentary rock which has very varying properties and colours. Often it has a proportion of limestone, just as some types of limestone have a proportion of sand. Some sandstone is very hard-wearing—such as Yorkstone and is used in paving, while other types are crumbly and need good detailing to prevent them weathering. Sandstone comes in many colours, from red to yellowish brown.

Limestone

Limestone and chalk are the sedimentary remains of countless living organisms from prehistoric seas. The sheer scale of the process of formation of these materials must surely rank as one of the ecological wonders of the natural world. Limestone is top-quality building stone, partly because it can be carved and cut with greater ease than most other building stones and partly because of its wonderful light honey colour. However it is attacked by acid rain and requires careful detailing to take account of this. At present it is extracted without due care in many parts of the country—especially in National Parks. If quarried as it used to be, with small-scale extraction as close as possible to the place of use, there would be less impact and each area would have its own local supply. It is a stone that needs to be used with greater care and respect.

Marble, a metamorphosed form of limestone, is often extracted with appalling environmental results, such as has happened in Italy. Britain has no indigenous supplies of marble, but certain decorative limestones, such as the shelly Purbeck Marble, are referred to as marble in the construction industry.

Slate

Slate is an excellent stone for shedding water and has the property of being easily split along the plane of its cleavage. High quality slates can be recycled several times before becoming weakened through weathering. The different qualities of slate need to be used in different ways to make the most of their characteristics. For instance, the strongest and best quality slates can be split quite thin and used where strength and lightness are paramount, whereas the weaker slates can be cut thicker or used in smaller sizes. Other slate is more appropriate for walling. If you have a slate roof and it is in need of renovation, it is worth finding out what you can about the type of slates used so that you can match them. It is also worth keeping a small stock of these slates for repairs as they arise. You can often tell where a particular slate has come from by its colour: for instance Westmoreland slate is generally green and that of North Wales is dark blue.

Vernacular character

Stone is one of the most important materials for giving an area a unique character. Think of sandstone used in the Cotswold villages, the limestone of the Dales and the flints of East Anglia. It is the use of these stones from local quarries that brings a great attractiveness to an area. However if we are ever to use stone in the ways it was used so successfully in the past, we need to redevelop the skills of quarriers and stonemasons. There are probably few more interesting trades.

If you have a house where the stone has been covered with render, consider removing a portion of the render to discover the quality of the stone behind. If it is of poor quality then you can re-render with an insulating layer. If, however, it is of reasonable quality it may be worth having it repaired and cleaned. The detailing of less durable stone is important to give it protection from the weather and prevent frost damage. If a building or wall has an obvious weathering defect, it could either be the result of a natural weakness of the stone or evidence of a need to provide the appropriate protection, such as an overhang or coping stones to help shed water away from the face.

Much newly quarried stone is used for hardcore and building up ground levels. If you require rubble or rough stone hardcore, consider using mine-stone, a by-product of mining. Artificial stone is a useful way of using up stone waste and is cheaper than stone; it is often simply concrete with a facing of stone dust mixed in.

Sand and gravel aggregates

There are large deposits of sand and gravel in Britain; huge quantities of these materials are extracted for the making of concrete. As with any raw material, it is important to draw up carefully thought-out ecological criteria for its extraction. We don't usually give much thought to this unless these materials are being taken from a much loved local beauty spot. Finding suitable deposits which do not result in environmental damage is becoming increasingly difficult, and is another reason for limiting the use of concrete where possible. However, with proper thought and landscaping, gravel pits can provide inland lakes for both recreation and wildlife. There may be potential for extending the proportion of aggregates that are dredged from the sea.

Earth and clay

Earth and clay are amongst the oldest building materials used by humans, and there is now renewed interest in their use. Experiments are now taking place around the country to regain the experience necessary for their use in a modern context. There are many different types of subsoil and clay, the nature of which depends on the many variables involved such as the type of rock from which the material was originally weathered, the different climatic conditions, and the past vegetation. These all have effects which create very different textures, structures and behaviour in use, and lime or clay was sometimes added to improve cohesion. As a result each area developed its own vernacular tradition, based on what was found to work.

As a building material, earth is abundant in most areas. It often requires little 'working' if dug from the best deposits. When dry it is strong yet flexible. If built thick enough then it has good thermal and sound-insulating properties. It cannot rot and, if protected from rain and detailed properly, can last indefinitely. Its embodied energy is almost negligible.

In Britain wattle and daub was in use as an infill material until the 18th century in parts of the country; many examples still exist often behind plaster rendering. What is not so well known is that rammed earth was a standard method of building in many villages until the beginning of this century. Its vernacular name and method of construction varied from one part of the country to another: cob, witchert, pise and clay lump were some of the names used. Some of the construction methods are now being updated, as the prospects for using this material in areas that have suitable deposits is good. Its durability is very dependent on careful detailing to protect it from rain and ground moisture. It requires a good roof overhang and a damp-proof course. On the inside of a building, rooms can be finished with plaster and, on the outside, whitewash or timber boarding. This type of construction might be particularly appropriate for a garden shed, garage or garden wall.

Rock asphalt

Rock asphalt is included here although it falls between the categories of natural stone and processed minerals. Occurring naturally in various parts of the world, it is the result of oil having seeped to the earth's surface and the volatile components having evaporated off. After heating to make it liquid, it can be laid as a flooring material or applied vertically or horizontally as damp-proofing. It is an extremely resilient and hardwearing material if used away from UV radiation. There are substitutes for rock asphalt such as those produced from oil or coal. The advantage of rock asphalt is that it is healthier to use inside the house as all the volatile compounds have long since evaporated off.

Raw mineral materials

A large proportion of the minerals mined are for processing into substances that often look very different from the raw product. There are three basic categories.
- Minerals such as sand, earth, clay, limestone and gypsum, which are used for making glass, bricks, lime, cement and plaster.
- Metallic ores such as iron ore or bauxite, which are processed into steel and aluminium.
- Coal, oil and natural gas, which are processed into all manner of chemicals, plastics and paints.

The problems that relate to the extraction of the raw material minerals are largely the same as for those minerals whose problems of over-exploitation, pollution and environmental damage have already been largely addressed. It is in the processing of some of these materials that even bigger problems arise; these are addressed in the next chapter.

Priorities for action

1. If you want to use stone, learn about your local stones and find out where the nearest working quarries are. It is difficult to predict how stone from a quarry will weather, so it is best to locate buildings known to have beensupplied from a particular quarry or, conversely, locate the quarry from which a favoured or particularly durable stone came.

2. If you are using aggregates, find out where your local source is and how serious the environmental problems are at the relevant quarries. This in turn might help you decide between using these aggregates or finding an alternative material.

3. If you have a requirement for rubble or hardcore, avoid using quarried stone, especially limestone. There are many alternatives available including minestone, the stone waste from the mining of coal and metallic ores. Better still, use recycled crushed concrete, broken bricks or material from road resurfacing.

4. Consider using earth as a building material if you have an appropriate use, such as for a structure in your garden.

Processed and synthetic materials

Somewhat like processed food, processed building materials are liable to be less healthy for us and for the planet than are natural ones. So what should we be watching out for with these types of materials ? There are three main concerns. The first is that many of the processes are themselves energy-intensive and polluting. The second is that of industrialisation itself: its tendency to become centralised and geared to quantity production often leaves in its wake a correspondingly large amount of damage to the environment. The third concern is the sheer number of new synthetic chemicals that new technologies have allowed us to produce—particularly the organic petrochemicals. Some of these are relatively harmless, but others are highly toxic and

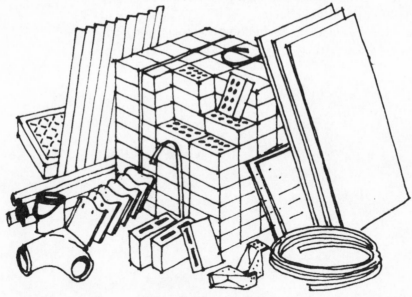

Processed materials.

often used unnecessarily as a substitute for natural products.

Entirely new materials are being produced every day. We cannot turn the clock back and disinvent them; however, in certain cases, we can wake up to their effects and if necessary ban or restrict their use.

The processes in this chapter are dealt with under three headings:
- Processing earth and rock
- Metal production
- Synthesising chemicals from coal, oil and gas

Processing earth and rock

The processing of earth and rock to produce bricks, plasters, cements and glass are amongst the oldest known human technologies. They vary considerably in their damage to the environment.

Bricks and Tiles

Bricks were first produced in a sun-baked version 6,000 years ago; they are perhaps the oldest processed building material. There have been many developments in fired-earth technology, and the same processes have been used to develop a whole range of components. In addition to brick there are floor tiles, wall tiles, roof tiles, firebricks and drainpipes of many different colours, properties and strengths, all of which have been developed from the same basic raw materials. Western Europe has made more use of this technology than any other region of the world.

In terms of aesthetics, hand-made bricks and tiles have a marvellous individuality of texture and character, ranging in colour through many shades of red, brown, yellow, grey and blue. It is the different firing methods, as well as the different constituents, that produce bricks of varying textures, colours and strength. Most of these components have been developed to a size that can be handled by one person.

To understand the ecological impact of these manufacturing processes, it is worth examining them in more detail. The raw materials come from three basic sources:

- Surface clays—typically from river and glacial deposits
- Shale clays—usually from surface outcrops
- Fireclays—often found deep underground beneath coal seams and so requiring mining—used only for special bricks

The different materials are blended to obtain the desired characteristics. Crushing and pounding of the harder materials such as the shales is usually necessary in order to provide a uniform consistency. There are also various other ingredients that are sometimes used, including calcium silicate sand and lime.

Before firing, water needs to be driven off through drying. Firing takes place at a temperature between 1, 600—2,000°C, depending on the type of clay used. This is the point at which most environmental damage can be done. At present kilns waste considerable quantities of heat because of the way the firing is carried out. This can be reduced by increasing the insulation and incorporating heat recovery techniques into the kilns. In addition, during firing, toxic gases and vapours which are both corrosive and polluting are often produced, depending on the raw materials used. This pollution can also be reduced by choosing different raw materials. Some manufacturers are doing just this.

If we require bricks for renovation, we should support those companies that are doing most to overcome these environmental problems. Companies concerned with ecological issues will also need to ensure that their extraction methods are not unduly destructive of habitats and that water tables are not adversely affected.

Lime

Lime is the basis of many building materials such as cement, plaster and certain types of bricks. It is also used in steel-making, in agriculture, for water treatment, and in innumerable industrial processes. It is also a constituent of paints and many industrial finishes. It is even used in scrubbing flue gas emissions of sulphur dioxide from power stations.

The actual manufacturing process which produces lime is a simple one. Limestone, chalk or shell deposits are heated in a kiln to a high temperature until carbon dioxide is given off, producing quicklime. This is then combined with water to produce slaked lime, which is the main constituent of lime plasters, lime mortar and lime washes. It gradually hardens over a period of time as it reabsorbs carbon dioxide from the atmosphere.

If we look at the overall ecological damage that is done through the production and use of lime, it is mainly the quantity of the material being extracted that is the biggest threat, as it is used in so many ways (see **LIMESTONE** on p.189). If we do not ensure that policies are changed then whole landscapes in areas of outstanding natural beauty are likely to be levelled. We therefore need to consider carefully our use of lime-based building materials, particularly those that use large quantities of lime in their manufacture.

Plasters

There are three different types of plaster.
- Lime-based plasters
- Gypsum-based plasters
- Cement-based plasters

Cement plasters are the strongest and also the most water resistant, and for this reason are chosen for external rendering. Although other plasters can be used externally, they need to be coated with a layer of protecting paint.

Gypsum is now the common basis for all modern interior plasters. Gypsum (calcium sulphate) is found naturally but is also produced artificially. Its main source is from natural deposits, which are mined in a similar fashion to limestone. In order to manufacture gypsum plaster, it is necessary to drive off some of the combined water in the gypsum. It is then ground to a powder.

When mixed with water it recombines and sets relatively quickly. Calcium sulphate is now also produced as a by-product of scrubbing the sulphur dioxide out of power station emissions. This source of calcium sulphate is now being used to supplement natural mined gypsum. The environmental problems that accompany the use of gypsum plasters are similar to that of any material that is extracted from the ground. However there are not yet the severe problems of ubiquitous use that occur with limestone.

Cement

The Romans used a type of slow-setting cement that they made from volcanic ash and lime; it was this type that was used until Portland cement was developed in the 18th century. Cement is a complex mixture of materials: there are many different blends, each with its own given name. In its simplest form lime and clay are mixed together and heated in an oven. The resultant material is then ground to form a powder which, on combining with water, sets to a hard and durable material; this powder forms the basis of cement mortar and concrete. Raw materials for modern cements use silica, alumina, iron oxides and different compounds of calcium, including gypsum.

There are a number of environmental problems associated with the production of cement. The very high temperatures ($1,300$—$1,500°C$) that are needed to burn the mixture require a large energy expenditure, and the gases and vapours that are given off in the process contain metals and oxides of sulphur. The chromates in cement are thought to be particularly dangerous. There are two other hazards to be aware of in using cement: breathing in the cement particles can cause silicosis of the lungs, and contact with wet cement can burn the skin.

It is yet another industry that uses limestone, but energy expenditure and pollution pose the biggest problem. There is much scope for developing a 'greener' cement.

Glass

Glass is certainly the most environmentally useful of the processed materials: its properties of being durable, transparent to light and reflective to infra-red gives it the ability to trap sunlight in way that no other material can do. Being almost totally inert, it is a healthy material both to have as part of a home and also for food and drink containers. It can easily be washed and reused or recycled by melting down.

The oldest glass found was made in Egypt around 2,500 BC. The Romans developed the technology and produced the first crude glassblowing. The Venetians rediscovered these techniques and developed them so that glass became available all over Europe. The industrial revolution produced changes to large-scale production and the development of drawn and plate glass.

What many people do not realise is that many substances can be used to form glass. The most widely used is silica (quartz-sand). Lime and sodium or potassium salts can be added to improve workability. These materials are fired at a temperature of 1,500—1,600°C at which point the glass is as fluid as water. There are three basic ways that glass is produced:

- vertical drawing—producing drawn glass
- flat sheet casting and rolling—producing patterned glass
- floating on molten metal—producing float glass

Many different compounds can be added to the glass to give it different properties or colours. Lead silicate produces glass of high refractive index and coloured glass is made by adding metals such as copper, nickel or cobalt to the melt.

Both sand and limestone are required for the making of glass, as is a large amount of energy. Owing to the detrimental ecological impact of its production, it is important that it is used as efficiently as possible. Glass has the added hazard that there is a danger of cuts from falling into it or through it; however, it can be made exceptionally strong by either laminating layers of glass together or toughening it by heat treatment. Safety glass is up to five times stronger, and if broken disintegrates into harmless fragments.

There are also various coatings that can be used to produce certain effects. Solar control glass is tinted to reduce glare and heat gain; and low-emissivity glass reflects heat back into the home, contributing to energy conservation.

Besides flat glass there are two insulating materials made from glass that are also relevant for energy saving:

- Glass fibre. It is used as an insulating material and reinforcement, made by blowing molten glass out of holes in a fast spinning drum, similar to making candyfloss. When made into quilts or batts the bonding agent used is usually urea-formaldehyde. A similar process is used for making rockwool, in which molten rock is used instead of molten glass. Slagwool is best of all because it is a by-product, and thus no new material has to be specially mined. It can also be spun directly from the furnace, thus saving energy.
- Foam glass is just what its name implies but is not commonly used because of its high cost. However, it is extremely durable and water resistant; this combination makes it relevant when considering the insulation of basements or underground buildings where there is the likelihood of the material remaining wet.

Ceramics

Coating fired clay products with a glaze (glass) is a method of preservation and decoration that is used for ceramic tiles, glazed tiles, glazed fireclay, vitreous china and vitreous enamels. Ceramic products, although having a high embodied energy, are generally ecological in that they are mostly extremely durable and can often be recycled. This recycling is facilitated if care is taken to fix them in the first place without too strong a bond.

Metal production

Most metals are environmentally destructive in their extraction. They also tend to be highly energy-intensive and polluting in their production, yet there are considerable differences in their damaging effects. We need to look carefully at the damage the extraction and production of these metals can do, and decide which metals should be avoided altogether and which limited as far as possible.

Of course there are some uses for metals for which there is no substitute, such as electrical and certain plumbing items, and fixings such as screws and nails. There are a number of metals that are now scarce, such as lead, tin, tungsten and zinc. However they are also poisonous, so maybe a reduced usage will be a blessing in disguise. Most metals are in fact toxic, including mercury, nickel, zinc, aluminium, silver, cobalt, cadmium, titanium, selenium and chromium. In general, the heavier the metal the more toxic it is likely to be. The exception is gold, which is stable and inert and thus harmless.

The most widely used metals in building are steel, copper, aluminium and lead, and these are dealt with next, in a little more detail.

Iron and steel

The raw materials for the production of iron are: iron ore (mainly oxides of iron), coke or charcoal, and limestone. 'Pig iron' is made by heating these materials together in a blast furnace. Steel is made by reheating the pig iron produced to purify it, and adding various materials to make different types of steel. Carbon is the critical ingredient which distinguishes steel from iron. Manganese, silicon and phosphorus are also added to give varying properties. Other alloys of iron include nickel, chromium, vanadium, tungsten, cobalt etc., which are used to produce stainless steel and specially strong steels or hard steels. From a domestic perspective, the most important of these alloys after steel is stainless steel, which contains 12–18% chromium and some nickel.

The main ecological effects of iron and steel production are the huge quantities of raw materials that are used and the energy consumed in the process. Perhaps the most damaging effect at present on a global scale is the destruction of vast areas of tropical rainforest that are being destroyed to produce charcoal for pig iron production in Brazil. More than 50% of this pig iron is being shipped to the West so that Brazil can repay the continuing interest on its foreign debts.

Copper

Copper is one of the few metals found in a free metallic state in nature, which accounts for its early use (around 8000 BC). Other metals in the same group are silver and gold. The raw materials for copper production are copper ores (mainly sulphides), native copper (its free state), or ores mixed with other ores of nickel, zinc or lead.

Processing consists of three stages:

1. The copper ore first has to be separated out by a process of grinding the different ores into a fine powder and mixing with water and reagents. This mixture is then violently agitated to produce a heavy froth which contains 95% of the copper ore.
2. After settling and drying, the resultant material is roasted to remove further impurities, including sulphur dioxide. This stage is the most polluting—producing the major ecological problem associated with this method of processing. Smelting reduces the large quantities of sulphur dioxide otherwise emitted.
3. Electrolytic methods are used for further refining.

At present each stage of the process is environmentally destructive in some way, whether it is the large amounts of water that are used or the sulphur dioxide that is released into the atmosphere from the sulphide ores.

Aluminium

Aluminium is the most abundant metal element in the earth's crust, but is surprisingly difficult to separate from its mineral compounds. It has therefore been a relative latecomer as a refined metal—it was not produced until 1827. The raw materials for its production are bauxite, which is rock containing about 50% aluminium oxide, sodium carbonate and limestone. The refining process is complicated, lengthy and highly energy-consuming. Smelting involves the aluminium oxide being reduced, using carbon heated in an electrolytic bath. Aluminium refining is the most energy-intensive process known for producing materials used in the construction industry, being dependent on electricity and using approximately 126 times the amount used for timber production. In its favour, aluminium is easily recycled, corrosion resistant and lightweight.

Lead

Lead is a highly toxic metal as are its soluble compounds. Its use in the building industry is being steadily reduced for this reason. From being the main material used for channelling water, whether on roofs or in gutters, downpipes and all internal pipework, it is being replaced mainly by copper and plastic.

It still is a useful material in roofing; however if rainwater is to be collected from a lead roof or lead guttering for use in the garden, then it would be wise to use this only on areas producing non-edible produce.

The production of lead is similar to that of copper. Galena (lead sulphide), which contains up to 86.6% lead, is the main raw material. However it also occurs dispersed with other minerals; the refining processes can be very complex, the aim being to recover traces of silver and gold along with other impurities such as arsenic, zinc and tin.

Other metals

Most other metals that are used in building are used as alloys or as pigments for plastics and paints.

The following metals are particularly toxic:

- chromium
- cadmium
- tin
- mercury

All these metals should be avoided where possible. Chromium is perhaps the most difficult metal to avoid as it is so much a part of existing kitchens and bathrooms, which often contain stainless steel sinks and chrome-plated fittings. Plating uses a small amount of the metal and aids durability. Cadmium is often used in batteries and these should be avoided where possible. If this proves impossible, used batteries containing heavy toxic metals should be disposed of as a toxic waste. Titanium is a metal that is being increasingly used in paint manufacture as a substitute for lead. However it is a difficult metal to extract and we should limit its use to essential requirements.

Chemicals from coal and oil

The third category of chemicals are those that result from the processing and synthesis of coal and oil. These form the constituents of a huge number of products.

Coke and its by-products

When coke is produced from coal, all the volatile compounds are driven off and these condense as coal tar, a mixture of hundreds of different compounds. This coal tar is refined into chemicals of different volatility by a process of distillation. From these chemicals a vast range of dyes, drugs, perfumes, explosives, antiseptics, plastics, resins, biocides, synthetic fibres, solvents etc. are produced. From this enormous range of chemicals, the building industry makes use of those that can be used for damp-proofing (pitch), timber treatment products (creosote), plastics and paints.

Apart from extraction damage, the environmental damage due to coke works and liquefaction refineries depends entirely on how well the production is run. There is no need for chemicals to leak into the atmosphere or local water supply, though this often occurs. What is perhaps the most damaging result for the environment is that many chemicals are produced as by-products which are then marketed as something that we need! Though many of them may be genuinely useful, a number are used only because it is cheaper to find a use for them than to dispose of them. We have little idea at present how many toxic chemicals have been foisted on us in this way.

Coke itself has many major industrial uses. Apart from the production of metals, it is used, for example, to reduce steam to hydrogen at high temperatures; and coke processed with lime produces calcium carbide. This product is used to produce acetylene, which in turn is used to produce plastics and synthetic rubbers.

Products from the distillation of crude petroleum

When crude petroleum is distilled in a similar way to coal tar, about half the number of compounds are produced—fuels, solvents, paraffins, lubricants and asphalt. The environmental effects of petroleum extraction are well known as they are considered newsworthy events. Perhaps again the most damaging effect on the environment is the way the petroleum industry has been run for maximum profit. Even relatively toxic by-products have to find a use even if to the detriment of human beings and their environment.

The oil industry is not an environmentally friendly one. Millions of tons of oil pollute the sea every year in completely preventable circumstances. Considerable air pollution occurs with the leaking of methane and the burning off of unwanted products that could be used. Oil well blow-outs and excessively high rates of extraction have had damaging environmental effects. The end products are sold so cheaply that we do not think twice about driving trivial distances for sheer convenience.

Plastics and paints are of special relevance to building and home-making, so we will look at them in some detail.

Plastics

By plastics we mean the synthetic plastics produced mainly from coal tar and petroleum. There are many types of plastic, which can be categorised into thermosets and thermoplastics, the latter accounting for about 80%. These can be melted by heat many times whereas thermosets, as the name implies, set only once. Most plastics are produced from petroleum and natural gas.

Thermoplastics used in the building industry include polyvinyl chloride (PVC), polyethylene, polypropylene (PP), acrylics, polystyrene, nylon, polycarbonate, polyvinyl butytal and cellulose. Thermosets include urea-formaldehyde, polyesters, polyurethane and silicones.

The ecological impact of plastics comes not only from the energy costs in producing them, but also from the fact that most plastics do not biodegrade like vegetable or animal matter, or corrode like metals. In addition, most plastics cannot easily be recycled or repaired. As a result, plastics end up being thrown away faster than any previous material and end up as a new form of pollution in the countryside and the sea. Some of them, particularly the thermoplastics, off-gas gradually, releasing plasticising agents and intermediate compounds. The amount of off-gassing depends to some extent on how well the plastic has been manufactured. Off-gassing increases with temperature, so it is a good idea to keep plastics away from heat sources in your home. If plastics catch fire, most burn easily, emitting thick, black toxic smoke. Plastics therefore need to be chosen with care, using natural alternatives where possible.

Paints and varnishes

Paints are a mixture of pigments to give colour, binder to bind them together, and a solvent to make the paint flow. In some cases all three pose a threat to the environment. Solvents are the most immediate pollutants as they are substances that are intended to be evaporated off. It has been estimated that (globally) over 500,000 tons of solvent are released into the atmosphere from paints each year. Inside your house these solvents can damage the health of whoever is decorating if there is insufficient ventilation. Amongst the solvents that are toxic are:

- white spirit
- toluene
- xylene
- benzene

In addition to the evaporation of solvents, paints can often continue to off-gas small quantities of toxic vapour for a considerable time (see TOXINS on p.115 and AIR on p.122). During the production of synthetic paints there is often a huge amount of chemical waste—a hidden source of pollution. With some paints, for each litre of paint several litres of pollutant are produced.

The present rate of paint use is unsustainable in many different ways, largely because of the many toxic chemicals that are involved. We need to look carefully at all the possible alternatives to the existing toxic mixtures that are sold as paint. There are, of course, important differences between the damaging effects of different paints but generally we cannot continue to use them as we are. Until more benign synthetic materials are developed, we should return to some of the more traditional ways of making paints from natural materials, such as linseed oil and turpentine.

Priorities for action

1. Choose bricks from companies that are working on improving their environmental performance.
2. Use cement sparingly.
3. Use glass products that increase energy conservation.
4. Choose durable and reliable metal appliances and equipment to avoid the need for replacement.
5. Choose lead-free pipes for plumbing and take care in removing and disposing of old flaking paint—it may contain lead.
6. Avoid aluminium products if a less energy-intensive material will perform the same task acceptably.
7. Avoid using plastic items when paper or wood products can serve the same purpose.
8. Use paints that use natural or benign constituents where possible. If in doubt refrain from redecorating and prolong the life of your present decorations. You can do this by cleaning your paintwork and using careful touching up instead.

Recycling

Our existing economy runs largely on a throughput of materials rather than recycling. Throughput means extracting whatever is considered necessary from the earth and forests, using what we want and discarding what is considered waste into the ecosystem (mainly the atmosphere and water systems). At present rates this cannot last very long before resource depletion, degradation and an overload of the natural cleansing systems set in. In the long run a sustainable way of running our affairs needs to be found, and the sooner we do this the less painful the future will be. What does sustainable mean in this context? In a simple way it means recycling our waste as an integral part of providing for our needs. Rather than doing it for a few limited materials, we need to aim to do it for everything. Every output should be a potential input for another process, as occurs in nature.

This requires much rethinking of the way that many of our goods are designed and manufactured. This is already happening in Germany, where the onus for recycling has been passed back to the originator of the waste. Every manufacturer now has to develop the methods of recycling as part of the design process. If there is no known way of recycling a particular material or component, then there should be no further production until a way has been found. This measure would have the effect of weeding out products that are not ecologically feasible.

Waste on the present scale is completely unprecedented. Past societies never generated the mountains of rubbish we produce every day. Nor did they plunder the earth's resources to the same extent: they simply did not have the technological means to do so. Even today there are rural societies, such as in Ladakh, that effectively generate no waste whatsoever.

How can we begin to achieve a reduction in throughput and an increase in recycling within our homes? If we think of our home as a materials system, much like its energy system (see p.38), products and materials bought become the inputs, and the materials and waste to be disposed of become the outputs. Both these inputs and outputs need to be reduced to a minimum. To help achieve this we need to re-use and recycle as much as possible within our homes. We need also to recycle the unwanted materials (outputs) in as many ways as may be necessary. This can be summarised below, in a way that gives the main headings for this chapter:

REDUCE—particularly items bought (inputs)
RE-USE—products etc. inside our homes
RECYCLE—unwanted materials etc. (outputs)

<u>REDUCE</u> inputs

How can we best reduce the inputs that really matter? We need particularly to look at three elements: the overall bulk of goods and materials that we are consuming; the ways in which goods and materials can be acquired from recycled sources; and the need to ensure that the remaining goods and materials are chosen using ecological criteria.

Limiting new inputs

There are many ways we can do this in our homes. We can do it partly by limiting the space we use in the first place. This will have an impact on the amount of materials needed for decorating and furnishing (let alone extending). The first chapter gives some ways we can make creative and efficient use of the space in our homes.

Another very important way in which we can reduce these inputs is to limit the wastage of building materials in any building work we carry out. Building materials wasted during construction account for about 20% of the total used. This wastage is cumulative at various stages of the building process. When specifying and ordering materials, inaccuracies commonly occur. Materials can also be spoiled during storage from traffic, poor stacking or moisture. The use of the materials also causes wastage through damage and off-cutting at the installation stage.

Our society encourages us to buy far more than we need. We ought therefore to find ways of clarifying what our real needs are, and focusing on meeting these rather than buying the many superfluous products that seem to demand our attention. In general we need to limit our consumerist tendencies and also improve the quantity of the goods we purchase.

Apart from buying less, we can share more. Rather than finding this an annoyance we could choose to do this with someone we like to interact with, thereby making the process more pleasurable. We can share DIY and garden tools such as shredders. Alternatively we can hire tools or machinery, if a hire shop is nearby.

Using recycled materials or appliances

Using recycled materials is another way we can reduce our throughput. At present only 1% of building materials is supplied from reclaimed materials. This is a minute fraction of what it should be. SALVO (see **ORGANISATIONS**) is an organisation aiming to increase the percentage to 5%. They produce a newsletter for both the building trade and designers.

We can use recycled materials in one of two ways: either as second-hand or salvaged materials, or as materials made from a recycled source.

Salvaged or reclaimed materials or products

Old materials have stood the test of time and are often far more attractive than their modern equivalents. Salvage yards of either architectural pieces or standard building materials are the places to seek out and visit if you are looking for these items. Some are very expensive, because of their scarcity value. However there can be others at very reasonable prices. It is often a matter of luck whether you find what you are looking for, but such places are well worth a visit because they are also like museums of old building materials. It is worth checking your local newspaper under 'Building products' or the Architectural Salvage Index (see **ORGANISATIONS**). Local demolition sites can also be a source of ready materials.

Recycled and reclaimed materials.

Products from recycled materials

If you cannot use a reclaimed material, then one from recycled materials is the next best thing. Examples of these are insulation made from recycled cellulose fibre from old newspaper, and underlay made from old tyres. It is also always worth asking your local builders' merchant if there is a recycled alternative to a particular product.

Materials from industrial byproducts

These are made from by-products such as pulverised fuel ash (PFA) from power stations, or slag from metal production. Pulverised fuel ash is made into lightweight aggregate, and slag is made into various products including slagwool—an insulation product similar to fibreglass. Minestone—waste from mining operations which can be used for hardcore—is another example of a use for a material that would otherwise be wasted.

Ecological criteria for new inputs

The question of what makes an ecological material has been developed at some length in the CRITERIA FOR MATERIALS SELECTION chapter at the beginning of this Part of the book. In terms of recycling, the following criteria are of particular relevance.

Renewable materials

These are materials that are recycled by nature—if managed sustainably they should be a priority choice. They are the products of living organisms—the most important being timber. Others are products from plants such as rubber, cork and cotton; and from animals, such as wool (see LIVING RESOURCES on p.176).

Durable materials

Durability results in a reduced frequency of replacement and thus reduces the throughput of materials. Many materials can either be designed or manufactured to be durable. Examples are bricks, glazed tiles, iron railings and indeed windows.

Recyclable

These are materials that are either reusable as components, such as slates; biodegradable, such as wallpaper; or recyclable by melting down, such as metals. If we buy materials that can easily be recycled, we solve future problems when they become outputs.

We can also reduce need for new inputs by repair or renovation:

<u>RE-USE</u> materials within our homes

How can we reduce the speed with which we use up materials in our home? Essentially, we can either make things last longer through renovation, repair and re-using items where possible; or we can recycle materials within our homes.

Renovation and repair

In the past, items were continuously repaired and maintained until they became worn out. Now we have the concept of the disposable item which always will look brand new because it is continually replaced.

We need to learn to look after our home and possessions with more care. We can do this in various ways, including learning how to use things properly and knowing when an appliance needs to be serviced. However, we need to be willing to repair rather than replace. This is true for almost any area of our home, whether it is touching up the paintwork or repairing a cupboard door. Sash window renovation provides perhaps the best example of all.

Sash window renovation

Many people have been persuaded by builders to replace their old sash windows for much uglier and often inferior standard plastic or timber ones. What is surprising about so many old sash windows is that the wood they are made from is often in very good condition. They may have a bit of rot in the bottom sill, but if the windows are taken out, stripped, repaired and reglazed with stepped double-glazing, you have a window which is often superior in performance and certainly superior in appearance to a modern plastic one. However, renovating these windows is a labour of love, which includes careful painting and draught-stripping, but it is also very satisfying. If you find the prospect too daunting to undertake yourself, there are firms which do an excellent job.

Recycle what you can within your home or garden

There are a number of ways that we can recycle materials within our homes. The most important is by using natural processes. For instance, through composting organic materials can be recycled into plants and vegetables; air can be partly purified using indoor plants and water can be treated outside in reedbeds.

Of course all the other ways of recycling are relevant here, for instance the recycling of basic materials such as timber, fabrics or hardcore and the recycling of component materials such as bricks, tiles and slates. These can all also become recyclable outputs.

<u>RECYCLE</u> Outputs

We live in a throw-away society. Nowhere is this better expressed than in the mountains of rubbish collected from our homes every week, most of which could be reused. This is well-known and yet, as a society, we are doing very little to reorganise these collection systems. In addition, a very large amount of re-usable building materials, including a great number of architectural antiques, are still being deposited as landfill. Each time 30 bricks are thrown away, the energy equivalent of a gallon of petrol goes up in carbon dioxide. We also throw away every day hundreds of tons of old, but perfectly re-usable, tropical hardwoods.

What is also quite extraordinary is that we often throw whole buildings away, with sometimes only the briefest of salvage operations to recover just the items that will 'make a quick buck'. We have to adjust our thinking to the idea of recycling everything in one way or another, and only throwing away as a last resort. The aim should be to minimise these rejected items and eventually eliminate them altogether.

In reality there is no such thing as throwing something away. Away to where? What we mean is that we want it away from us. This only means that this waste is deposited either nearer someone else or into an increasingly sensitive ecosystem. What is more, it will often cycle back to us. Throw batteries in the bin and their mercury, lead, cadmium, sooner or later, show up in your water.

Demolition

When undertaking demolition work, it is important to dismantle the construction carefully in order to conserve as many of the materials as possible. Whether this is removing slates from a roof, taking brickwork apart, or extracting timber studs from a partition, a small amount of extra care delivers a large increase in the quality of the reclaimed materials.

Different ways of recycling outputs

The following are ways of recycling:
- Direct recycling of basic materials such as earth or timber.
- Recycling of reclaimed building components such as bricks, tiles and slates.
- Recycling by the melting down of such materials as glass, metal and some plastics (see next section).
- Recycling by repair or renovation.
- Recycling by down-grading—for instance waste bricks, concrete and stone become hardcore.
- Recycling through biodegradation or composting of organic matter.

Common to all recycling is the need to categorise and keep separate all the different items. Mixing materials nearly always leads to a downgrading. In the case of your own home, if you are having a builder do some work discuss with him how different materials can be reclaimed and kept separate, rather than throwing everything together into a skip.

Passing on your own unwanted building materials

There are various possible ways to find a home for your unwanted building materials. You can stack them up neatly and advertise in the local paper or let your neighbours know. Alternatively you can ring a local merchant: there are some traders who make it their business to know exactly who wants what. The range of materials and components that can be sold locally is much wider than generally imagined. There is usually someone requiring hardcore, second hand timber or offcuts. For anything more obscure like an old chimneypot or pieces of moulding you should look for a specialist in the field. Your Yellow Pages or 'SALVO' (see ORGANISATIONS) should be able to help. If it is of particular significance it can be registered on the Architectural Salvage Index mentioned earlier.

Recycling materials to be melted down

There are three basic types of material that are recycled by melting down— scrap metal, glass and plastics:

- Scrap metal—This is probably the best developed of all recycling collection systems in Britain, probably because it has been going for so long. There are undoubtedly huge improvements that could be made to the efficiency of the system, but so long as the different metals to be recycled are separate, you can even make some money.
- Glass—bottle banks are now ubiquitous. All that has to happen is for it to be made compulsory to use them. Those who manage them do not however encourage their use for broken panes of glass—sometimes there is even a notice to this effect. Small quantities can be broken up (with care!); larger quantities (if the council do not collect them) should be referred to your nearest glass recycling agent.
- Plastics—These are an increasing problem as the plastics industry in Britain is only under voluntary agreement to organise recycling schemes. Most plastic banks become overloaded very quickly and all manner of plastics get mixed together. Central and local government need to act together to enforce a proper system of plastics recycling that deals with the full range. For our part we need to clean and separate our plastics into different categories. You will need to find out which plastics are recycled locally.

Dealing with toxic wastes

Advice on what to do with toxic substances such as pesticides, solvents, lead paint scrapings, asbestos or old batteries is dealt with in TOXINS AND POLLUTANTS (see p.115). As a first step, contact the person responsible in your local council.

Priorities for action

1. Reduce the quantities of materials that you actually use as far as possible. When undertaking any sort of renovation work, calculate carefully the quantities required so that there is as little waste as possible.
2. Obtain and use recycled materials and reclaimed components wherever possible.
3. Reuse and recycle materials within your own home to reduce throughput. This means working on self-sufficiency. Growing, repairing, reusing, etc. all help.
4. Find ways of recycling all the materials and components that you don't want. This means reducing the amount of 'rubbish' that is actually 'thrown away' (taken by the bin-men) and finding outlets for all recyclable materials.

Part Five

Further Information

Assessing Your Home
Saving Money and Energy
Information on Products and Services
Organizations
Classification of Toxins
Recommended Further Reading
References and Bibliography
Index

Assessing Your Home: A Checklist

Most properties will require careful assessment, whether you are buying a new house or have been living in the same house for twenty years. In the former case, you will be looking to compare one house with another. In the latter, you may feel you know your house well. However, looking at your house afresh, the chapter headings may have given you new insight. Needless to say, there will also be a big difference in approach required between a house that needs substantial rebuilding and a house completed to almost the last detail. The former will require re-evaluating in almost every aspect, whereas the latter demands attention mainly to finishes, services and appliances.

The first four survey checklists follow the structure of the book; they deal with space, energy, health and materials. Structure and water penetration are included under 'materials'. Where appropriate, practical comments are added. It should be emphasised that a full appreciation of the questions and comments cannot be gained unless you read the main text! Where this is particularly important the relevant chapter is referred to.

Space survey

Internal space

Can all functions and activities be accommodated efficiently?
To determine this, it is worth making a complete list of all the essential activities that you wish to accommodate. If you then place these in order of priority, you can assign them to different rooms and odd spaces in the house, starting with the top priority and working down the list. Once you have allocated space to as many of the functions and activities as possible, you may then need to consider how more than one function can be accommodated in one space. With this method you should soon be able to assess if the total space available is sufficient for your needs.

Conversions

If there is insufficient space, is an internal conversion possible?
Look to see if there is any unused space, within the confines of the existing walls and roof, that can be converted. You may well be able to squeeze some extra space from odd corners in your roof space, basement or outhouses. Look also at the different options for storage (p.9-11) since this can save space and use it more efficiently.

External space

How far can recreational, self-sufficiency and habitat needs be accommodated?
Treat the outdoor space similarly to the internal space in terms of prioritising activities and functions, taking careful account of orientation and shading.

Extending ecologically

Is there the potential for an ecological extension?
After reading the chapter on **INTERNAL SPACE AND CONVERSIONS**, you should be clear

as to whether all your internal space is allocated efficiently. If more accommodation is essential, look at the possibilities of extending:

- Underground
- Above ground as a sun room or conservatory
- By putting up a separate building

Energy survey

Sources of energy

Check gas supply installation. Is it where you need it?
List all places in the house where electricity is used for heating.
How easy is it for these electrical functions to be changed to a more ecological form of heating?

Draught-proofing

Check all doors and windows for fit.
Burn a stick of incense and watch the smoke to check for draughts.
Read the chapter on **DRAUGHT-PROOFING AND VENTILATING** to understand the likely causes of air pressure differences, such as the stack effect; and check the prevailing wind direction.

Insulating

Identify the extent of existing insulation.
Check the insulation of your loft, walls, windows (double-glazing) and under the floors. Where is it nonexistent, and where is it insufficient?

Is external wall insulation feasible?
If there is no cavity wall, there is the question of whether to place insulation on the outside of the building (the most energy-efficient position) or the inside (likely to be internally disrupting).

Space heating

Check space heating installation for energy efficiency.
Read the chapter on **SPACE HEATING** and check:

- Boiler
- Pipes and radiators
- Controls

Check the whole arrangement for space heating in terms of the overall efficiency of the system (see pp.75-79).

How old is your boiler?
Will it be too big once the additional insulation is installed?
If it is a gas boiler, is it a condensing type?
Find out whether the heating controls work, and whether they are outdated (there have been significant improvements in heating control design in recent years). You may be able to work out yourself roughly what is likely to need changing or adding; find out what you can before you call in an expert.

Lighting

What proportion of the house is lit by energy-efficient lighting?
How well is daylight used?

Appliances

Which appliances are essential to you?
How energy-efficient are your existing appliances?
It helps to make a list of the appliances you have; refer to the APPLIANCES chapter to help you to estimate how energy-efficient they are likely to be. This is often difficult to do precisely, but, for example, if you look at the time it takes for the machine to carry out a cycle and how many times it heats up water, you will get a rough idea.

Solar energy

Check orientation of the house and passive solar potential.
Is there a suitable site for a future solar panel?
The SOLAR ENERGY chapter will help you decide how high a priority the installation of one or more solar panels should be, given your particular circumstances and home energy needs.

Health Survey

Sources of toxins

List the potential sources of toxins around the house.
Identify any serious sources and consider what would be necessary to replace them (see TOXINS chapter).

Water

Taste the drinking water supply after running the tap.
This can give you an instant (but very subjective) judgement as to quality. Also, find out what you can from your local water company. The WATER chapter indicates what to look for. Particularly important are (a) acidity leading to possible lead contamination, and (b) the quantity of chlorine, and the likelihood of it reacting with water from wetlands. Check nitrate contamination levels and whether fluoride is added.

Radon

Check if you are in an area where there is a radon risk.
If you live in a risk area, you should have your house tested for radon contamination (see pages 142-144). This is particularly true if you are planning to use any basement areas as living space. It is a good idea to check that all your ventilation bricks are free of obstructions, and that there is a good flow of air through your underfloor cavities, both to vent any radon and also to keep these spaces dry, to help to protect the floor joists.

Sunlight and health

How easy is access to the outside to sit in the sun?

Sound

Check sound-proofing between floors and partitions.
If near a busy road or airport, are your windows sound-proofed?

You will know where the weak points are in your sound-proofing between rooms (refer to the SOUND chapter for ways of dealing with this). If you are assessing a house for prospective purchase, you will have to carry out two tests: walking with heavy shoes, and shouting.

In both these cases one person should be making the noise in one room, and the other should be listening in an adjacent one. Remember that furnishings will absorb sound so in an empty house every noise will sound much louder.

Plants

Where are the places that plants could be located?
There may be an opportunity for adding extra space for plants, such as an oriel or bay window, or even an add-on greenhouse.

Materials

Here we are concerned mainly with the durability of materials in terms of the integrity of the structure and any degradation caused by water penetration. The traditional structural survey and a water penetration survey are therefore relevant in relation to the long-term viability of the dwelling. You may also wish to survey the property with the intention of replacing worn materials with more ecologically sound ones.

Structural survey

Making a structural survey is a professional's job, but it is possible for most people to get a good impression of the state of the structure of their own, or a possible future house, by making a number of simple observations and local enquiries. These are as follows:

Information from the neighbourhood

Look for recurring problems in adjacent housing.
For example if the same house design is repeated, there may be a common problem, for instance around bay windows, or cracks on the facade that may not be immediately obvious in the house you are concerned with.

If the house is terraced, ask neighbours if there are any party wall problems.
Ask if you can look at the party wall from their side, and find out whether they have had to do any repairs on the wall. This will give you an idea as to whether there are any serious problems that are hidden by redecoration.

Ask neighbours about the ground and any foundation problems.
You will need to find out what the subsoil is like. The worst problems occur with clay, peaty soil and built-up land.

Information from your own observations

Look for differential settlement, cracks, and walls and sills out of true.
You may not see directly, but it is often possible to tell whether differential settlement has taken place, for instance where a heavier part of the building has sunk further than an adjacent wall. This sometimes happens with a heavy chimney breast or a load-bearing wall adjacent to a lighter partition. If you think there is any external wall that is leaning, you can check this with a plumb-line.

Check the strength of floors.
There is a simple way of checking how strong your floors are: jump up and down in the middle of the room, with someone observing the deflection. This is a somewhat crude method, but if there is a serious weakness you will both see it and feel it. You may not have to do more than go on your toes to feel a deflection.

Check for unsupported partitions.
Partitions that are built on a floor with no support from below can often be a cause of weakness, which may show up in cracks where the partition meets the walls or ceiling.

Check the roof structure for woodworm and rot.
Roof structures are normally timber; the most likely problems are rot and woodworm. If the roof is lined, you may not have easy access to the main timbers, but they are often exposed near the eaves, which is the most common area where problems occur. Woodworm may show up with little powder deposits—on cobwebs, for instance; the difficulty is to know how serious the problem is. It may be a very minor problem, or the woodworm may no longer be there. (Try to solve the problem without resorting to the use of highly toxic chemicals.)

Check the structural integrity of the stairs.
As with floors, you can often tell there is a problem by using them. If there is a weakness and the underside is accessible, use a torch to observe what happens when someone uses them. (There are various ways of improving the state of stairs from underneath, both to strengthen them and to stop them creaking if necessary.)

Check the joint between any adjunct to the main structure such as a bay window or extension.
There are often problems if a bay window was added after the main structure was built, because the foundations of the bay may be insufficient, or not tied into the main structure. If you perceive a problem, you will need to identify whether it is a matter of the bay window simply moving independently of the main house structure, or whether there is a more serious structural defect relating to the size of the opening in the main structure.

Check carefully any balconies or other cantilevered structures.
The structural integrity of balconies depends on a well-designed solution to the cantilever. Often a cantilever becomes weakened and the structure can become unsafe. This can usually be identified by looking at the jointing with the main structure. (If there is any uncertainty, it would be worthwhile getting a structural engineer to check it.)

Check parapets and chimneys.
These are both elements that can become weakened by weathering, with the danger that they may be blown down in a storm. This is not uncommon, and may become more frequent if the severity of storms increases due to climate change. Examine these structures carefully, both up on the roof (if you can), and also through a pair of binoculars on a bright day. Chimneys can also become decayed from the inside by the action of smoke, which can crumble the mortar, the brickwork, or both. If it is possible to get near the chimney or parapet, tapping with a hammer can often help to tell how solid the structure is.

Water penetration

The next most important area to be concerned about is the shedding of rainwater. Water leaking into the house or on to the wall of the house over a period of time is likely to cause decay of materials, particularly timber.

Check the roof and flashings.
A careful look at the roof from the outside, with binoculars if possible, will usually show up any defects such as loose or missing slates, dislodged ridge tiles or loose flashings. If the chimney pots are not capped, this can be a source of damp. If it is possible to inspect the roof from the inside, look for chinks of light and places where there are stains due to water penetration.

Flat roofs cause particular problems. Sometimes a leak does not show through on the underside until it has been there for some time. Have a careful look at the state of the covering material and, if you think there may be a problem, obtain advice from someone experienced with flat roofs.

Check gutters and down spouts.
These are prone to blocking and leaking, which can pour water onto the brickwork for long periods with no one noticing. The serious damage this can do makes it important to take the opportunity when it is raining to observe how well both gutters and down spouts are functioning.

Check for signs of rising damp.
Rising damp comes from moisture in the ground. There may be signs of moisture along the skirting board area of a wall on the ground floor, which has caused the decorations to deteriorate. It is important to distinguish between rising damp and condensation: the former can be distinguished by white sulphate salt deposits. (Condensation should cease to be a problem if the house in question is insulated and properly ventilated.) If there *are* signs of rising damp, this often means there is no damp-proof course.

Damp-proof courses (DPCs) are meant to prevent damp from rising above them. So finding your damp course and checking that it runs all the way round the building is the next important step to take. If there is no DPC it is still possible that, because of the way the building is designed, there will be little problem with rising damp. (However, although other improvements may be far more important, it is often a condition of a mortgage from a building society that you should install a DPC.)

Damp in a basement or cellar is to be expected as this was one of the reasons cellars were built in the first place: to evaporate away rising moisture from the ground. If you want to solve any problems here, it requires someone experienced in damp-proofing basements.

Check gullies and drains.
It is worth checking that there are no blockages and that water flows through without impediment. To test that they are free-flowing, lift the access cover of each gully and pour down a bucket of water.

Saving Money and Energy

If you have extremely limited resources, the most important measures to take are those that will lead to the most immediate savings, in the hopes of subsequently taking further steps that will have longer-term impact. Energy-efficient light bulbs, for instance, will begin saving you money from the moment you install them. Since many utilities are beginning to subsidize the purchase of new compact fluorescent bulbs, it is a good idea to check with your local electric company for discounts before buying from a retailer.

Low-Income Weatherization Programs, funded in part by the U.S. Department of Energy, are available in most states and free to eligible applicants. Certified energy auditors compile data using computerized worksheets to determine which energy conservation measures are most effective for your home. Based upon these results, the homeowner can modify or convert inefficient heating systems, add insulation, draught-proof, and take other measures—again at no charge for those who qualify.

It may also be worthwhile to have your home inspected for a home energy rating by Energy Rated Homes (ERH). This nonprofit organization offers real incentives to homeowners of very efficient as well as very wasteful houses. If you are purchasing or thinking of refinancing the mortgage on a home that is already energy-efficient, a Four Star rating will entitle you to better mortgage rates. If, on the other hand, ERH issues you a rating below Four Star, you can capitalize by financing tax-deductible improvements through your mortgage. The rating process also provides comprehensive information on cost-effective improvements and energy-cost projections. For details, and to see if your state has its own ERH inspection service, contact:

Energy Rated Homes of America
100 Main Street
Little Rock, Arkansas 72201
(501) 771-2299

Reading relevant books and articles need not cost you anything if you have a good public library, and such information can lead to major energy savings. The more you are able to see your house as a complete energy system, the more choices you can make to save money.

Another excellent source of free advice is the technical staff at Real Goods Trading Corporation of Ukiah, California. Accessible via phone, mail, or electronic mail, these specialists can help you assess your energy conservation needs, select components, and design your own energy system.

Real Goods Trading Corporation, Technical Service Dept.
555 Leslie Street
Ukiah, CA 95482
(800) 762-7325 or (707) 468-9292 outside the U.S.
www.realgoods.com
email: realgood@realgoods.com

Information on Products and Services

Your local architectural or building society or a well-stocked public library should have access to the following and similar reference works. Since the following source-books and periodicals are often resource-intensive and costly to buy or subscribe to, using them in a reference setting is not only cheaper, but more ecologically sound.

AIA Environmental Resource Guide Subscription. Published quarterly by the American Institute of Architects, 1735 New York Ave., NW, Washington, DC 20006; (800) 365-2724. www.aia.org.
This periodical focuses primarily on materials and large-scale building issues, but it often contains current information on retrofits as well.

Real Goods Solar Living Sourcebook. 10th Edition. Ukiah, CA: Real Goods Trading Corporation, 1994. Available from Real Goods (see **ORGANIZATIONS** section).
A thick, informative resource catalog devoted to all aspects of independent living. Features sections on energy-efficient lighting, water conservation and purification, photovoltaics and solar electric systems for "off-the-grid" living, and much more. One of the best product guides available, with descriptions and schematic drawings as well as prices and ordering information. Products for average homeowners, as well as for renewable-energy gurus.

Consumer Guide to Home Energy Savings by Alex Wilson and John Morrill. Berkeley, CA.: American Council for an Energy-Efficient Economy (ACEEE), annual. Available through bookstores or from ACEEE (See **ORGANIZATIONS** section).
An invaluable book for anyone about to purchase a new appliance or just wanting to save on energy bills. The guide lists every kind of home appliance, from dishwasher to furnace, by brand name and evaluates them based on energy efficiency. Also includes information on weatherizing or upgrading your home.

The Efficient House Sourcebook by Robert Sardinsky and the staff of the Rocky Mountain Institute. Snowmass, CO: RMI, 1992. Available from Real Goods or the Rocky Mountain Institute (see **ORGANIZATIONS** section).
Offers reviews of selected books and a directory of organizations devoted to "home-scale resource efficiency." Lists and critiques the periodicals, books, schools, organiza-tions, and agencies that deal with all aspects of resource-efficient house design, construction, retrofitting, and more.

Environmental Building News, 122 Birge St., Suite 30, Brattleboro, VT 05301; (802) 257-7300; (802) 257-7304 (fax); www.ebuild.com; ebn@ebuild.com
This bimonthly newsletter is primarily concerned with new environmentally safe materials and construction techniques, but many concepts apply equally well to reno-vations. Written by and for building professionals, but impressively well informed and ecologically sound.

The Interior Concerns Resource Guide and Newsletter by Victoria Schomer, ASID.
Mill Valley, CA: Interior Concerns Publications, annual. Fax: (415) 388-8322.
The *Resource Guide* lists building and decorating materials and resources, organized by
type. There are entries on environmentally correct woods, recycled nails, hypoaller-
genic paints and carpeting, etc. The emphasis is on nonpulluted interiors made with
sustainable materials. The *Resource Guide* is updated annually. Between updates,
Interior Concerns covers current news as well as theory and practice in its bimonthly
newsletter.

Safe Home Digest, Lloyd Publishing, Inc., 24 East Ave., Suite 1300, New Canaan,
CT 06840; (203) 966-2099.
A monthly newsletter featuring articles on current environmental health issues. Also
publishes the *Safe Home Resource Guide*, a comprehensive sourcebook of healthy
products and services listing everything from natural paints and glues to alternative
insulation and flooring.

*The Sourcebook for Sustainable Design: A Guide to Environmentally Responsible
Building Materials and Processes*, ed. by Andrew St. John, AIA. Boston: Architects
for Social Responsibility, 1992. Available from Boston Society of Architects, 52
Broad St., Boston, MA 02109; (617) 951-1433, ext. 221; (617) 951-0845 (fax);
www.architects.org; bsarch@architect.org.
Organized by conventional architectural divisions that relate to specific aspects of
building (for instance, Division 7 is titled "Thermal and Moisture Protection"), it is
possible to find appropriate practices and resources quickly.

Stewart's Green Line Environmental Directory, ed. by Mary Lou Stewart. Delta, BC:
Stewart, annual.
Lists more than 12,000 environmentally inspiring products, technologies, services,
and organizations within 850 categories, including Environmental Architects and
Biological Pest Control. Available through Real Goods Trading Corporation (see
ORGANIZATIONS section).

The following books also contain lists of products and suppliers:
Ecologue by Bruce Anderson
The Green Consumer by John Elkington, et al.
The Green Home Handbook by Gillian Martlew and Shelley Silver
Home Ecology by Karen Christensen
The Natural House Book by David Pearson
Your Natural Home by Janet Marinelli and Paul Bierman-Lytle
For further details on these books, turn to **RECOMMENDED FURTHER READING** and
REFERENCES AND BIBLIOGRAPHY.

Organizations and Businesses Providing Information, Products and Services

Local energy providers, state and county energy officers, and housing authorities often have good, relevant information on materials, procedures, appliances, and strategies for ecological renovation. Since these agencies are numerous and varied, we have made no effort to list them here among the following organizations and companies. Although practitioners of ecological building and retrofitting are generally a conscientious and honest lot, we suggest you seek local references and check with your local Better Business Bureau before retaining any consulting or renovation services.

The Alliance to Save Energy, 1200 18th Street, NW, Suite 900, Washington, DC 20036; phone (202) 857-0666; (202) 331-9588 (fax); www.ase.org; ase@info.org. Promotes energy efficiency through research, demonstration projects, lobbying, and education.

American Council for an Energy Efficient Economy (ACEEE), 2140 Shattuck Ave., #202, Berkeley, CA 94704; (510) 549-9914. Publishes the annual *Consumer Guide to Home Energy Savings* (see Information on Products and Services), as well as a Guide to Energy Efficient Office Equipment, which evaluates everything from fax machines to laser printers.

American Institute of Architects (AIA), 1735 New York Ave., NW; Washington, DC 20006; (202) 626-7300; www.aia.org. Publishes the annual *AIA Environmental Resource Guide Subscription* (see **INFORMATION** section) and can make referrals to local architects who use environmental design principles.

American Solar Energy Society, 2400 Central Avenue, Unit G-1, Boulder, CO 80301; (303)443-3130; (303) 443-3212 (fax); www.ase.org/solar; ases@ases.org. The U.S. branch of the International Solar Energy Society. ASES publishes *Solar Today*, a bimonthly magazine that examines current developments in renewable energy technologies on both the home and utility scale.

American Wind Energy Association, 777 North Capitol Ave., NE, Washington, DC 20002; (202) 383-2500; (202) 408-8536 (fax).

Audubon Society, 950 Third Ave., New York, NY 10022; (212) 979-3000; (212) 353-0347 (fax).

Center for Resourceful Building Technology, P.O. Box 100, Missoula, MT 59806; (406) 549-7678; (406)549-4100 (fax); www.montana.com/crbt; crbt@montana.com. A clearinghouse of information for home builders, architects, and consumers. Developers of the ReCRAFT '90 demonstration house, which incorporated a wide variety of recycled building materials in its construction. Publishes the *Guide to Resource-Efficient Building Elements*, which lists dozens of suppliers of products made from post-consumer waste.

The Consumer Federation of America, 1424 16th St., NW, Suite 604, Washington, DC 20036; (202) 265-7989; www.consumerfed.org; cfa@essential.org. An important public-interest lobbying organization.

Energy Efficient Building Association (EEBA), P.O. Box 2307, Eagan, MN 55122-0307; (651) 994-1536; www.eeba.org; info@eeba.org.

Environmental Construction Outfitters, 44 Crosby St., New York, NY 10012; (212) 334-9659 or (800) 238-5008. Distributor of more than 1,000 ecologically safe building products, including nontoxic paints, salvage lumber, "green" appliances, etc. Clients include homeowners, builders, and architects.

Greenpeace USA, 1436 U St., NW, Washington, DC 20009; (202) 462-4507; www.greenpeaceusa.org; international: www.greenpeace.org.

The Healthy House Institute, 430 N. Sewell Road, Bloomington, IN 47408; (812) 332-5073; www.hhinst.com; healthy@bloomington.in.us. An independent resource center focusing on indoor air-quality issues. Sells books such as *Healthy House Building: A Design and Construction Guide*, and offers professional consultation by phone at hourly rates.

National Center for Appropriate Technology (NCAT), P.O. Box 4000, Butte, MT 59702; (406) 494-4572; (406) 494-2905 (fax). Provides up-to-date information and assistance on energy conservation and renewable energy.

The Owner Builder Center, 1250 Addison St., Suite 209, Berkeley, CA 94702; (510) 848-6860.

Radon Testing Corporation of America (RTC), 2 Hayes St., Elmsford, NY 10523; (800) 457-2366; (914) 345-8546 (fax); www.rtca.com. Distributes and analyzes charcoal canister radon tests for homeowners.

Real Goods Trading Corporation, 555 Leslie Street, Ukiah, CA 95482; (800) 762-7325 (orders); (707) 468-9292 (business office); (707) 468-9214 (technical information); www.realgoods.com; realgood@realgoods.com. A mail-order catalog company dedicated to renewable energy, ecologically appropriate technologies, and products for independent living. Publishes the exhaustive and useful *Solar Living Sourcebook* (see Information on Products and Services) and sponsors seminars through its Institute for Independent Living.

Rocky Mountain Institute, 1739 Snowmass Creek Rd., Snowmass, CO 81654-9199; (970) 927-3851; (970) 927-9178 (fax); www.rmi.org; outreach@rmi.org. A think tank that studies all aspects of energy efficiency and its implications for society and the future. Publishes *The Efficient House Sourcebook* (see **INFORMATION ON PRODUCTS AND SERVICES**) and *Practical Home Energy Savings* (see **REFERENCES AND BIBLIOGRAPHY**).

Sierra Club, 730 Polk St., San Francisco, CA 94109; (415) 776-2211.

Union of Concerned Scientists, 2 Brattle Square, Cambridge, MA 02238-9105; (617) 547-5552; (617) 864-9405 (fax); www.ucsusa.org; ucs@ucsusa.org. Organization of scientists and citizens concerned with the impact of advanced technology on society and the global ecosystem.

U.S. Department of Energy (DOE), 1000 Independence Ave., SW, Washington, DC 20585; (202) 586-5000; www.doe.gov. Questions and referrals concerning all aspects of energy, including government policies, grants, and rebates for home energy conservation measures and renewable energy systems.

U.S. Department of Housing and Urban Development (HUD), 451 7th St., SW, Washington, DC 20024; (202) 708-1422. Information on home testing services and suppliers and federal grant programs for homeowners.

U.S. Environmental Protection Agency (EPA), 401 M St., SW, Washington, DC 20460; (202) 260-2090; www.epa.gov. Questions and referral on issues of suspected toxins in the home, including pesticides (identification and disposal of banned products) and radon testing (recommended laboratories and testing services). It is usually best to contact your regional EPA office from the following list:

Boston: EPA, 1 Congress St., Suite 1100, Boston, MA 02114-2037; (617) 918-1111.

New York: EPA, 290 Broadway, New York, NY 10007-1866; (212) 637-3000.

Philadelphia: EPA, 1650 Arch St., Philadelphia, PA 19103-2029; (215) 814-2090.

Atlanta: EPA, Sam Nunn Federal Building, 61 Forsyth St. SW, Atlanta, GA 30303; (404) 562-9900.

Chicago: EPA, 77 West Jackson Blvd., Chicago, IL 60604; (312) 353-2205.

Kansas City: EPA, 726 Minnesota Ave., Kansas City, KS 66101; (913) 551-7000.

Dallas: EPA, 1445 Ross Ave., Dallas, TX 75202; (800) 887-6063.

Denver: EPA, 999 18th St. Suite 500, Denver, CO 80202-2466; (800) 227-8917.

San Francisco: EPA, 75 Hawthorne St., San Francisco, CA 94105; (415) 744-1500.

Seattle: EPA, 1200 Sixth Ave., Seattle, WA 98101; (206) 553-1200.

Classification of Toxins

The following list is intended as a quick reference guide to toxins. In the main text, there is a classification of toxins according to their source. Here, the headings are mainly chemical, with a small number of source headings at the end. Besides the common names, the relevant technical names are also given for those readers who would like to know them. The listing will help you to gain an understanding of the categories of chemicals under which toxins are more likely to occur.

Much of this information has been extracted from *Buildings and Health—The Rosehaugh Guide* (see **REFERENCES**), and those wishing to obtain more detailed information should consult this book. For a simpler treatment of the subject, *The Green Home Handbook* is more accessible. The categories below are, of course, not mutually exclusive.

METALS AND THEIR COMPOUNDS
Aluminum, cadmium, chromium compounds (chromates particularly), lead, mercury and metalloids—arsenic and antimony.

HALOGENS AND THEIR COMPOUNDS
Chlorine, bromine and fluorine and compounds such as bromides and fluorides.

MINERALS
Natural mineral asbestos fibres: chrysotile (white asbestos), amosite (brown asbestos), crocidolite (blue asbestos), tremolite (fine white asbestos) and actinolite (green asbestos). Other natural minerals: crystalline silica dust, talcum and quartz dust.

ACIDS, ALKALIS, OXIDISING AND REDUCING AGENTS
Alkalis such as ammonia, sodium hydroxide (caustic soda). *Acids* such as formic acid. *Oxidising agents* such as ozone and hydrogen peroxide and *reducing agents* such as carbon monoxide.

INORGANIC GASES
Nitrogen dioxide, sulphur dioxide and hydrogen sulphide.

ORGANIC COMPOUNDS
Volatile organic solvents: acetone, amyl acetate, benzine, carbon tetrachloride, perchloroethylene, toluene diisocyanate, trichloroethane (methyl chloroform), trichchloroethylene and xylene. *Aromatic hydrocarbons*: benzine, toluene, xylene, ethylbenzine, trimethylbenzine, styrene, napthalene. *Chlorinated hydrocarbons*: methylene chloride, 1,1,1-trichlorethane, trichloroethylene, chloroform, tetrachloroethylene, chlorobenzine and dichlorobenzine.

Certain families of organic compounds: aldehydes, phenols, benzenes, alkanes, polycyclic aromatic hydrocarbons.

PLASTICS & RUBBERS

Although most plastics in use cannot be classed as toxins, the quantities that are used mean that allergic effects from contact and off-gassing of minute quantities of toxic vapours can have a cumulative effect. The following may cause problems: polyester, poly vinyl chloride (PVC), polyurethane resin and foam (PU), urea formaldehyde (UF) and plasticizers e.g. dioctyl phthalate. Also *synthetic rubbers*: nitrile rubbers

LIVING ORGANISMS AND THEIR PRODUCTS

Disease organisms such as viruses and bacteria, certain moulds and fungal spores, some pollens, certain poisonous plants, dusts from particular species of timber and terpenes extracted from pine: alpha-penene, limonene.

PESTICIDES

Insecticides—bendiocarb, borates, cypermetblin, dichlorvos, dieldrin, permethrin, synthetic pyretbroids, lindane, pentaclorophenol PCB, sodium dichromate, tributyltinoxide (TBTO) and chlordane. *Fungicides*: arsenic pentoxide, copper sulphate, creosote and dichlorfluanid. *Herbicides*: diquat, atrazine.

RADIOACTIVE GASES

Radon and thoron.

TOXINS COMMONLY FOUND IN WATER

Aluminium, coliforms, copper, faecal streptococci, lead, manganese, nitrates, organochlorine compounds, poly-aromatic hydrocarbons, pesticides, sodium and trihalomethanes.

TOXINS COMMONLY FOUND IN INDOOR AIR

Carbon dioxide, formaldehyde, particulates, nitric oxide, nitrogen dioxide, ammonia, acrolien, ozone, sulphur dioxide, toluene and styrene.

PRODUCTS OF COMBUSTION

Benzopyrenes, carbon monoxide, nitrogen dioxide, sulphur dioxide, hydrogen cyanide, ammonia, acrolein and isocyanates.

Recommended Further Reading

The following books cover particular subjects in greater depth and detail than has been possible in *Eco-Renovation*. Additional titles, including books on more specific subjects, can be found in the REFERENCES AND BIBLIOGRAPHY section.

General
Beyond the Limits: Confronting Global Collapse, Envisioning a Sustainable Future by Donella H. Meadows, Dennis L. Meadows, and Jørgen Randers (White River Junction, VT: Chelsea Green Publishing, 1992). A scientific look at current trends in human population growth and resource consumption. Data-rich, but written for intelligent lay-people.

Gaia: An Atlas of Planet Management, ed. by Norman Myers (New York: Doubleday, 1992). Gives an overview of global imbalances in many types of resources, discusses how these resources are currently managed, and suggests options for improvement.

Healing Gaia: A New Prescription for the Living Planet by James Lovelock (New York: Crow, 1991). For anyone wanting a scientific but accessible introduction to the interactions and cycles of our global ecosystem.

Home Ecology: Simple and Practical Ways to Green Your Home by Karen Christensen (Golden, CO: Fulcrum Publishing, 1990). A useful and readable book covering a wide range of the subjects introduced in *Eco-Renovation*.

The Natural House Book: Creating a Healthy, Harmonious, and Ecologically Sound Home Environment (New York: Simon & Schuster, 1989). The result of many people's input. International in scope, and the best overview to date on the constituents of ecological building. Don't be mesmerized by the glossy color photographs and miss the practical points discussed in the small print.

Space
Designing and Maintaining Your Edible Landscape Naturally by Robert Kourik (Santa Rosa, CA: Metamorphic Press, 1986). The most complete guide to managing your external space for both maximum yield and attractiveness.

Green Architecture: Design for an Energy-Conscious Future by Brenda and Robert Vale (Boston: Bulfinch Press, 1991). An attractively produced book that essentially analyzes the design of new buildings. Most of the material is universally applicable.

The Integral Urban House: Self-Reliant Living in the City by Helga Olkowski et al./Farallones Institute (San Francisco: Sierra Club Books, 1979). Now a bit dated, but shows a serious attempt to transform a house in the inner suburbs into a self-sufficient home; highly informative and comprehensive.

Sunwings: The Harrowsmith Guide to Solar Addition Architecture by Merilyn Mohr (Camden East, OH: Camden House Publishing, 1985; dist. by Firefly Books). A wealth of practical examples of what has been done to optimize solar energy by people extending their homes.

A Shelter Sketchbook: Timeless Building Solutions by John S. Taylor (White River Junction, VT: Chelsea Green Publishing, 1997). This collection of pen-and-ink sketches depicts and explains more than six hundred elegantly simple and practical structures created by centuries of anonymous builders.

Energy

Energy Efficient Building, ed. by Susan Roaf and Mary Hancock (New York: Halsted Press, 1992). A collection of current technical papers on different aspects of energy conservation and efficiency in building.

The Independent Home: Living Well with Power from the Sun, Wind, and Water by Michael Potts (White River Junction, VT: Chelsea Green Publishing, 1993). A detailed overview of modern home energy systems, featuring profiles of renewable energy pioneers living throughout the United States.

The Passive Solar House: Using Solar Design to Heat & Cool Your Home by James Kachadorian (White River Junction, VT: Chelsea Green Publishing, 1997). This book offers a technique for building homes that heat and cool themselves in a wide range of different climates, using ordinary materials and methods familiar to all building contractors and many do-it-yourselfers.

Who Owns the Sun? People, Politics, and the Struggle for a Solar Economy by Daniel M. Berman and John T. O'Connor (White River Junction, VT: Chelsea Green Publishing, 1996). Narrated against a backdrop of diminishing fossil fuels, environmental degradation, avaricious corporations, and worldwide competition for natural resources, this book shows how existing solar technologies combined with local management present logical remedies for our energy gluttony.

Health

Buildings and Health: The Rosehaugh Guide, ed. by Steve Curwell (RIBA Publications, 1990). A British book. The most complete technical guide to date on most aspects of buildings and health. Expensive.

Day Light Robbery: The Importance of Sunlight to Health by Dr. Damien Downing (London: Arrow Books, 1988). This is the one book I know that spells out why sunlight is so important to human health. Some of it may be a bit over the top, but it does help balance the present-day arguments against exposure to sunlight due to excess ultraviolet radiation.

The Green Home Handbook: A Guide to Safe and Healthy Living in a Toxic World by Gillian Martlew and Shelley Silver (London: Fontana Press, 1991). An easy-to-follow analysis of the toxins in our homes and the ways in which we can combat them.

Healing Environments: Your Guide to Indoor Well-Being by Carol Venolia (Berkeley, CA: Celestial Arts, 1988). A more spiritual approach to this subject, by an American architect.

Your Home, Your Health, and Well-Being by David Rousseau, W.J. Rea, and Jean Enwright (Berkeley, CA: Ten Speed Press, 1988). A thorough, practical guide to all kinds of health-related factors in the home. Chock-full of useful information; one of the best books on the subject.

Materials

Environmental by Design: A Sourcebook of Environmentally Aware Material Choices by Kim Leclair and David Rousseau (Point Roberts, WA: Hartley & Marks, 1993). A highly informative and easy-to-use handbook. For other reference works listing environmentally safe building materials, refer to the **INFORMATION** section.

The Good Wood Guide by Friends of the Earth UK (London: Friends of the Earth, 1988). A simple guide to assessing the ecological soundness of timber supplies and finding out which kinds of wood to avoid. For more information, contact Friends of the Earth UK, 26-28 Underwood St., Islington, London N1 7JQ; telephone (071) 490-1555.

Green Design: A Guide to the Environmental Impact of Building Materials by Avril Fox and Robin Murrell (Architecture Design and Technology Press, 1989). An alphabetically organized arrangement of ecological considerations, relating mostly to building materials.

The Pattern of English Building by Alec Clifton-Taylor (London: Faber and Faber, 1987). This book gives some fascinating insights into the history of building materials.

The Straw Bale House by Athena Swentzell Steen, Bill Steen, and David Bainbridge, with David Eisenberg (White River Junction, VT: Chelsea Green Publishers, 1994). This elegant book shows how straw bale construction is an exceptionally durable and inexpensive way to build a beautiful house. An excellent resource for professionals and lay-people alike.

References & Bibliography

General
Ecologue: The Environmental Catalogue and Consumer's Guide for a Safe Earth by Bruce N. Anderson (Englewood Cliffs, NJ: Prentice-Hall, 1990).
The Green Consumer: A Guide for the Environmentally Aware by John Elkington, et al. (New York: Penguin Books, 1990).
Imperiled Planet: Restoring Our Endangered Ecosystems by Edward Goldsmith, et. al. (Cambridge, MA: MIT Press, 1992).
Real Goods Solar Living Sourcebook, 10th edition ed. by John Schaeffer (Ukiah, CA: Real Goods, 1999).
Shopping for a Better Environment by Lawrence Tasaday with Katherine Stevensons (Deephaven, MN: Meadowbrook Press, 1991).

Space: General
Alternative Housebuilding by Mike McClintock (New York: Sterling Publishing, 1989).
The Autonomous House: Design and Planning for Self-Sufficiency by Brenda and Robert Vale (New York: Universe Publishing, 1977).
Designing Your Natural House by Charles G. Woods and Malcolm Wells (New York: Van Nostrand Reinhold, 1992).
Greenhome: Planning and Building the Environmentally Advanced House by Wayne Grady (Camden East, ON: Camden House Publishing, 1993).
The Naturally Elegant Home by Janet Marinelli with Robert Kourik (Boston: Little, Brown, 1992).
Places of the Soul: Architecture and Environmental Design as a Healing Art by Christopher Day (The Aquarian Press, 1990; dist. by Thorsons, San Francisco).
Shelter, ed. by Lloyd Kahn (Bolinas, CA: Shelter Publications, 1973).
Shelter II, ed. by Lloyd Kahn (Bolinas, CA: Shelter Publications, 1979).
The Timeless Way of Building by Christopher Alexander (New York: Oxford University Press, 1979).

Internal Space
The Compact House Book, ed. by Don Metz (Pownal, VT: Garden Way Publishing, 1983).
New Compact House Designs, ed. by Don Metz (Pownal, VT: Storey Publishing, 1991).
Nomadic Furniture 1 by James Hennessey and Victor Papanek (New York: Pantheon, 1973).
Nomadic Furniture 2 by J. Hennessey and V. Papanek (New York: Warner Books, 1980).
Small Houses by Lila Gault and Jeffrey Weiss (New York: Warner Books, 1980).
The Smart Kitchen by David Goldbeck (Woodstock, NY: Ceres Press, 1989).

Conversion and Renovation
Renovating Brick Houses by Phillip J. Decker and T. Newell Decker (Pownal, VT: Garden Way Publishing, 1990).
Creating Space without Adding On by Jack P. Jones (Blue Ridge Summit, PA: TAB Books, 1993).
Renovating Old Houses by George Nash (Newtown, CT: Taunton Press, 1992).
Renovation: A Complete Guide by Michael W. Litchfield (Englewood Cliffs, NJ: Prentice-Hall, 1990).
Skylights: The Definitive Guide to Planning, Installing, and Maintaining Skylights and Natural Light Systems by Tom Jensen (Philadelphia: Running Press, 1983).
Solar Projects for Under $500 by Mary Twitchell (Pownal, VT: Garden Way Publishing, 1985).

External Space
America's Neighborhood Bats by Merlin D. Tuttle (Austin, TX: University of Texas Press, 1988).
Building a Healthy Lawn by Stuart Franklin (Pownal, VT: Garden Way Publishing, 1988).
Earth Ponds: The Country Pond Maker's Guide to Building, Maintenance and Restoration by Tim Matson (Woodstock, VT: Countryman Press, 1991).
Forest Gardening by Robert A. de J. Hart (White River Junction, VT: Chelsea Green Publishing).
Let It Rot: The Gardener's Guide to Composting by Stu Campbell (Pownal, VT: Garden Way Publishing, 1990).
Permaculture: A Practical Guide for a Sustainable Future by Bill Mollison (Cedar Crest, NM: Permaculture Services International, 1992).
Permaculture Way: Practical Ways to Create a Self-Sustaining World by Graham Bell (San Francisco: Thorsons, 1992).
Root Cellaring: Natural Cold Storage of Fruits and Vegetables by Mike and Nancy Bubel (Pownal, VT: Garden Way Publishing, 1991).

Extensions
Building with Heart by Christopher Day (Totnes, Devon: Green Books, 1990).
Designing Houses: An Illustrated Guide to Building Your Own Home by Les Walker and Jeff Milstein (New York: Overlook Press, 1979).
Gentle Architecture by Malcolm Wells (New York: McGraw-Hill, 1991).
How to Build an Underground House by Malcolm Wells (Brewster, MA: Malcolm Wells, 1990).
The Self-Build Book by John Broome and Brian Richardson (Totnes, Devon: Green Books, 1991; dist by Chelsea Green Publishing, White River Junction, VT).
Sunspaces: New Vistas for Living and Growing (Pownal, VT: Garden Way Publishing, 1987).

Energy
Counting the Cost of Global Warming by John Broome (White Horse Press, 1992; dist. by Paul & Co., Concord, MA).
Cut Your Electric Bills in Half by Ralph J. Herbert (Emmaus, PA: Rodale Press, 1986).
547 Tips for Saving Energy in Your Home by Roger E. Albright (Pownal, VT: Storey Publishing, 1991).

Energy Conservation in Housing by David L. Meinert (New York: Vantage Press, 1990).

Massachusetts Audubon Energy Booklets by Massachusetts Audubon Society (1982-91).

Practical Home Energy Savings by David Bill and the staff of the Rocky Mountain Institute (Snowmass, CO: Rocky Mountain Institute, n.d.).

Renewable Energy: A Concise Guide to Green Alternatives by Jennifer Carless (New York: Walker & Co., 1993).

Renewable Energy: Sources for Fuels and Electricity, ed. by Thomas B. Johansson, et al. (Washington, DC: Island Press, 1993).

Draught-proofing, Ventilating and Insulating
Home Insulation by Harry Yost (Pownal, VT: Storey Publishing, 1991).

Superhouse by Don Metz (Charlotte, VT: Garden Way Publishing, 1981).

Space and Water Heating
The Harrowsmith Country Life Guide to Wood Heat by Dirk Thomas (Charlotte, VT: Camden House Publishing, 1992).

The New Woodburner's Handbook by Stephen Bushway (Pownal, VT: Storey Publishing, 1992).

Passive Solar Energy: The Homeowner's Guide to Natural Heating and Cooling by Bruce N. Anderson and Malcolm Wells (Amherst, NH: Brick House Publishing, 1993).

Thermal Shutters and Shades: Over 100 Schemes for Reducing Heat Loss through Windows by William A. Shurcliff (Andover, MA: Brick House Publishing, 1980).

Solar Energy
The Consumer Guide to Solar Energy by Scott Sklar and Kenneth Sheinkopf (Chicago: Bonus Books, 1991).

The Fuel Savers: A Kit of Solar Ideas for Your Home, Apartment, or Business, ed. by Bruce N. Anderson (Lafayette, CA: Morning Sun Press, 1991).

The New Solar Home Book by Bruce N. Anderson and Michael Riordan (Andover, MA: Brick House Publishing, 1987).

The Solar Electric House by Steven J. Strong with William G. Scheller (Still River, MA: Sustainability Press, 1993).

Health
Design for Good Acoustics and Noise Control by J.E. Moore (Port Washington, NY: Scholium International, 1988).

Health and Light by John N. Ott (Greenwich, CT: Devin-Adair, 1988).

The Healthy House by John Bower (Secaucus, NJ: Carol Publishing, 1991).

Well Body, Well Earth by Mike Samuels and Hal Zina Bennett (San Francisco: Sierra Club Books, 1983).

Toxins, Air and Water
Health Homes in a Toxic World: Preventing, Identifying, and Eliminating Hidden Health Hazards in Your Home by Maury M. Breecher and Shirley Linde (New York: John Wiley & Sons, 1992).

The Home Water Supply; How to Find, Filter, Store, and Conserve It by Stu Campbell (Pownal, VT: Garden Way Publishing, 1983).
The Nontoxic Home and Office by Debra Lynn Dadd (Los Angeles: Jeremy P. Tarcher, 1992).
Nontoxic, Natural & Earthwise by Debra Lynn Dadd (Los Angeles: Jeremy P. Tarcher, 1990).
Radon: A Homeowner's Guide to Detection and Control by Bernard Cohen (New York: Avon Books, 1989).
Radon: The Invisible Hazard by Michael LaFavore (Emmaus, PA: Rodale Press, 1987).
Why Your House May Endanger Your Health by Alfred V. Zamm and Robert Gannon (New York: Simon and Schuster, 1980).

Plants

The Complete Indoor Gardener, ed. by Michael Wright and Dennis Brown (New York: Random House, 1979).
The Small Garden by John Brookes (Avenal, NJ: Outlet Book Co., 1991).
The Solar Greenhouse Book, ed. by James McCullagh (Emmaus, PA: Rodale Press, 1978).

Materials

Building with Junk: A Guide to Home Building and Remodeling Using Recycled Materials by Jim Broadstreet (Port Townsend, WA: Loompanics, 1990).
Hazardous Building Materials by S.R. Curwell and C.G. March (London: Spon, 1986).
Illustrated Handbook of Vernacular Architecture by R.W. Brunskill (London: Faber and Faber, 1971).
Recycling: A Windstar EarthPulse Handbook by Susan Hassol and Beth Richman (Snowmass, CO: Windstar Foundation, 1989).
Salvaged Treasures: Designing and Building with Architectural Salvage by Michael W. Litchfield (New York: Van Nostrand Reinhold, 1983).

Index

Chelsea Green

Sustainable living has many facets. Chelsea Green's celebration of the sustainable arts has led us to publish trend-setting books about organic gardening, solar electricity and renewable energy, innovative building techniques, regenerative forestry, local and bioregional democracy, and whole foods. The company's published works, while intensely practical, are also entertaining and inspirational, demonstrating that an ecological approach to life is consistent with producing beautiful, eloquent, and useful books, videos, and audio cassettes.

For more information about Chelsea Green, or to request a free catalog, call toll-free (800) 639-4099, or write to us at P.O. Box 428, White River Junction, Vermont 05001. Visit our Web site at www.chelseagreen.com.

Chelsea Green's titles include:

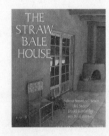 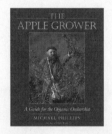 

The Straw Bale House

The Independent Home: Living Well with Power from the Sun, Wind, and Water

Independent Builder: Designing & Building a House Your Own Way

The Rammed Earth House

The Passive Solar House

The Sauna

Wind Power for Home & Business

The Solar Living Sourcebook

A Shelter Sketchbook

Mortgage-Free!

Hammer. Nail. Wood.

The Apple Grower

The Flower Farmer

Passport to Gardening: A Sourcebook for the 21st-Century

The New Organic Grower

Four-Season Harvest

Solar Gardening

Straight-Ahead Organic

The Contrary Farmer

The Contrary Farmer's Invitation to Gardening

Forest Gardening

Whole Foods Companion

Simple Food for the Good Life

The Bread Builder

Gaviotas: A Village to Reinvent the World

Who Owns the Sun?

Global Spin: The Corporate Assault on Environmentalism

Hemp Horizons

A Patch of Eden

A Place in the Sun

Renewables Are Ready

Beyond the Limits

Loving and Leaving the Good Life

The Man Who Planted Trees

The Northern Forest

Scott Nearing: The Making of a Homesteader